ON THE STUDY AND DIFFICULTIES OF MATHEMATICS

AUGUSTUS DE MORGAN

HAWK PRESS

Published by

Hawk Press
4836/24, Ansari Road, Daryaganj
New Delhi – 110 002
Phones : 9643330713, +91-11-23278618, +91-11-35676207
E-mail : thehawkpress@gmail.com
www.thehawkpress.com

ISBN : 978-93-93971-80-7

EDITOR'S NOTE.

No apology is needed for the publication of the present new edition of *The Study and Difficulties of Mathematics*,—a characteristic production of one of the most eminent and luminous of English mathematical writers of the present century. De Morgan, though taking higher rank as an original inquirer than either Huxley or Tyndall, was the peer and lineal precursor of these great expositors of science, and he applied to his lifelong task an historical equipment and a psychological insight which have not yet borne their full educational fruit. And nowhere have these distinguished qualities been displayed to greater advantage than in the present work, which was conceived and written with the full natural freedom, and with all the fire, of youthful genius. For the contents and purpose of the book the reader may be referred to the Author's Preface. The work still contains points (notable among them is its insistence on the study of logic), which are insufficiently emphasised, or slurred, by elementary treatises; while the freshness and naturalness of its point of view contrasts strongly with the mechanical character of the common text-books. Elementary instructors and students cannot fail to profit by the general loftiness of its tone and the sound tenor of its instructions.

The original treatise, which was published by the Society for the Diffusion of Useful Knowledge and bears the date of 1831, is now practically inaccessible, and is marred by numerous errata and typographical solecisms, from which, it is hoped, the present edition is free. References to the remaining mathematical text-books of the Society for the Diffusion of Useful Knowledge now out of print have either been omitted or supplemented by the mention of more modern works. The few notes which have been added are mainly bibliographical in character, and refer, for instance, to modern treatises on logic, algebra, the philosophy of mathematics, and pangeometry. For the portrait and autograph signature of De Morgan, which graces the page opposite the title, The Open Court Publishing Company is indebted to the courtesy of Principal David Eugene Smith, of the State Normal School at Brockport, N. Y.

THOMAS J. MCCORMACK

LA SALLE, Ill., Nov. 1, 1898.

AUTHOR'S PREFACE.

In compiling the following pages, my object has been to notice particularly several points in the principles of algebra and geometry, which have not obtained their due importance in our elementary works on these sciences. There are two classes of men who might be benefited by a work of this kind, viz., teachers of the elements, who have hitherto confined their pupils to the working of rules, without demonstration, and students, who, having acquired some knowledge under this system, find their further progress checked by the insufficiency of their previous methods and attainments. To such it must be an irksome task to recommence their studies entirely; I have therefore placed before them, by itself, the part which has been omitted in their mathematical education, presuming throughout in my reader such a knowledge of the rules of algebra, and the theorems of Euclid, as is usually obtained in schools.

It is needless to say that those who have the advantage of University education will not find more in this treatise than a little thought would enable them to collect from the best works now in use [1831], both at Cambridge and Oxford. Nor do I pretend to settle the many disputed points on which I have necessarily been obliged to treat. The perusal of the opinions of an individual, offered simply as such, may excite many to become inquirers, who would otherwise have been workers of rules and followers of dogmas. They may not ultimately coincide in the views promulgated by the work which first drew their attention, but the benefit which they will derive from it is not the less on that account. I am not, however, responsible for the contents of this treatise, further than for the manner in which they are presented, as most of the opinions here maintained have been found in the writings of eminent mathematicians.

It has been my endeavor to avoid entering into the purely metaphysical part of the difficulties of algebra. The student is, in my opinion, little the better for such discussions, though he may derive such conviction of the truth of results by deduction from particular cases, as no *à priori* reasoning can give to a beginner. In treating, therefore, on the negative sign, on impossible quantities, and on fractions of the form $\frac{0}{0}$, etc., I have followed the method adopted by several of the most esteemed continental writers, of referring the explanation to some particular problem, and showing how to gain the same from any other. Those who admit such expressions as $-a$, $\sqrt{-a}$, $\frac{0}{0}$, etc., have never produced any clearer method; while those who call them absurdities, and would reject them altogether, must, I think, be forced to admit the fact that in algebra the different species of contradictions in problems are attended with distinct absurdities, resulting from them as necessarily as different numerical results from different numerical data. This being granted, the whole of the ninth chapter of this work may be considered as an inquiry into the nature of the different misconceptions, which give rise to the various expressions above alluded to. To this view of the question I have leaned, finding no other so satisfactory to my own mind.

The number of mathematical students, increased as it has been of late years, would be much augmented if those who hold the highest rank in science would condescend to give more effective assistance in clearing the elements of the difficulties which they present. If any one claiming that title should think my attempt obscure or erroneous, he must share the blame with me, since it is through his neglect that I have been enabled to avail myself of an opportunity to perform a task which I would gladly have seen confided to more skilful hands.

AUGUSTUS DE MORGAN.

CONTENTS.

CHAPTER I.

INTRODUCTORY REMARKS ON THE NATURE AND OBJECTS OF MATHEMATICS.

THE OBJECT of this Treatise is—(1) To point out to the student of Mathematics, who has not the advantage of a tutor, the course of study which it is most advisable that he should follow, the extent to which he should pursue one part of the science before he commences another, and to direct him as to the sort of applications which he should make. (2) To treat fully of the various points which involve difficulties and which are apt to be misunderstood by beginners, and to describe at length the nature without going into the routine of the operations.

No person commences the study of mathematics without soon discovering that it is of a very different nature from those to which he has been accustomed. The pursuits to which the mind is usually directed before entering on the sciences of algebra and geometry, are such as languages and history, etc. Of these, neither appears to have any affinity with mathematics; yet, in order to see the difference which exists between these studies,—for instance, history and geometry,—it will be useful to ask how we come by knowledge in each. Suppose, for example, we feel certain of a fact related in history, such as the murder of Cæsar, whence did we derive the certainty? how came we to feel sure of the general truth of the circumstances of the narrative? The ready answer to this question will be, that we have not absolute certainty upon this point; but that we have the relation of historians, men of credit, who lived and published their accounts in the very time of which they write; that succeeding ages have received those accounts as true, and that succeeding historians have backed them with a mass of circumstantial evidence which makes it the most improbable thing in the world that the account, or any material part of it, should be false. This is perfectly correct, nor can there be the slightest objection to believing the whole narration upon such grounds; nay, our minds are so constituted, that, upon our knowledge of these arguments, we cannot help believing, in spite of ourselves. But this brings us to the point to which we wish to come; we believe that Cæsar was assassinated by Brutus and his friends, not because there is any absurdity in supposing the contrary, since every one must allow that there is just a possibility that the event never happened: not because we can show that it must necessarily have been that, at a particular day, at a particular place, a successful adventurer must have been murdered in the manner described, but because our evidence of the fact is such, that, if we apply the notions of evidence which every-day experience justifies us in entertaining, we feel that the improbability of the contrary compels us to take refuge in the belief of the fact; and, if we allow that there is still a possibility of its falsehood, it is because this supposition does not involve absolute absurdity, but only extreme improbability.

In mathematics the case is wholly different. It is true that the facts asserted in these sciences are of a nature totally distinct from those of history; so much so, that a comparison of the evidence of the two may almost excite a smile. But if it be remembered that acute reasoners, in every branch of learning, have acknowledged the use, we might almost say the necessity, of a mathematical education, it must be admitted that the points of connexion between these pursuits and others are worth attending to. They are the more so, because there is a mistake into which several have fallen, and have deceived others, and perhaps themselves, by clothing some false reasoning in what they called a mathematical dress, imagining that, by the application of mathematical symbols to their subject, they secured mathematical argument. This could not have happened if they had possessed a knowledge of the bounds within which the empire of mathematics is contained. That empire is sufficiently wide, and might have been better known, had the time which has been wasted in aggressions upon the domains of others, been spent in exploring the immense tracts which are yet untrodden.

We have said that the nature of mathematical demonstration is totally different from all other, and the difference consists in this—that, instead of showing the contrary of the proposition asserted to be only improbable, it proves it at once to be absurd and impossible. This is done by showing that the contrary of the proposition which is asserted is in direct contradiction to some extremely evident fact, of the truth of which our eyes and hands convince us. In geometry, of the principles alluded to, those which are most commonly used are—

I. If a magnitude be divided into parts, the whole is greater than either of those parts.

II. Two straight lines cannot inclose a space.

III. Through one point only one straight line can be drawn, which never meets another straight line, or which is *parallel* to it.

It is on such principles as these that the whole of geometry is founded, and the demonstration of every proposition consists in proving the contrary of it to be inconsistent with one of these. Thus, in Euclid, Book I., Prop. 4, it is shown that two triangles which have two sides and the included angle respectively equal are equal in all respects, by proving that, if they are not equal, two straight lines will inclose a space, which is impossible. In other treatises on geometry, the same thing is proved in the same way, only the self-evident truth asserted sometimes differs in form from that of Euclid, but may be deduced from it, thus—

Two straight lines which pass through the same two points must either inclose a space, or coincide and be one and the same line, but they cannot inclose a space, therefore they must coincide. Either of these propositions being granted, the other follows immediately; it is, therefore, immaterial which of them we use. We shall return to this subject in treating specially of the first principles of geometry.

Such being the nature of mathematical demonstration, what we have before asserted is evident, that our assurance of a geometrical truth is of a nature wholly distinct from that which we can by any means obtain of a fact in history or an asserted truth of metaphysics. In reality, our senses are our first mathematical instructors; they furnish us with notions which we cannot trace any further or represent in any other way than by using single words, which every one understands. Of this nature are the ideas to which we attach the terms number, one, two, three, etc., point, straight line, surface; all of which, let them be ever so much explained, can never be made any clearer than they are already to a child of ten years old.

But, besides this, our senses also furnish us with the means of reasoning on the things which we call by these names, in the shape of incontrovertible propositions, such as have been already cited, on which, if any remark is made by the beginner in mathematics, it will probably be, that from such absurd truisms as "the whole is greater than its part," no useful result can possibly be derived, and that we might as well expect to make use of "two and two make four." This observation, which is common enough in the mouths of those who are commencing geometry, is the result of a little pride, which does not quite like the humble operation of beginning at the beginning, and is rather shocked at being supposed to want such elementary information. But it is wanted, nevertheless; the lowest steps of a ladder are as useful as the highest. Now, the most common reflection on the nature of the propositions referred to will convince us of their truth. But they must be presented to the understanding, and reflected on by it, since, simple as they are, it must be a mind of a very superior cast which could by itself embody these axioms, and proceed from them only one step in the road pointed out in any treatise on geometry.

But, although there is no study which presents so simple a beginning as that of geometry, there is none in which difficulties grow more rapidly as we proceed, and what may appear at first rather paradoxical, the more acute the student the more serious will the impediments in the way of his progress appear. This necessarily follows in a science which consists of reasoning from the very commencement, for it is evident that every student will feel a claim to have his objections answered, not by authority, but by argument, and that the intelligent student will perceive more readily than another the force of an objection and the obscurity arising from an unexplained difficulty, as the greater is the ordinary light the more will occasional darkness be felt. To remove some of these difficulties is the principal object of this Treatise.

We shall now make a few remarks on the advantages to be derived from the study of mathematics, considered both as a discipline for the mind and a key to the attainment of other sciences. It is admitted by all that a finished or even a competent reasoner is not the work of nature alone; the experience of every day makes it evident that education develops faculties which would otherwise never have manifested their existence. It is, therefore, as necessary to *learn to reason* before we can expect to be able to reason, as it is to learn to swim or fence, in order to attain either of those arts. Now, something must be reasoned upon, it matters not much what it is, provided that it can be reasoned upon with certainty. The properties of mind or matter, or the study of languages, mathematics, or natural history, may be chosen for this purpose. Now, of all these, it is desirable to choose the one which admits of the reasoning being verified, that is, in which we can find out by other means, such as measurement and ocular demonstration of all sorts, whether the results are true or not. When the guiding property of the loadstone was first ascertained, and it was necessary to learn how to use this new discovery, and to find out how far it might be relied on, it would have been thought advisable to make many passages between ports that were well known before attempting a voyage of discovery. So it is with our reasoning faculties: it is desirable that their powers should be exerted upon objects of such a nature, that we can tell by other means whether the results which we obtain are true or false, and this before it is safe to trust entirely to reason. Now the mathematics are peculiarly well adapted for this purpose, on the following grounds:

1. Every term is distinctly explained, and has but one meaning, and it is rarely that two words are employed to mean the same thing.

2. The first principles are self-evident, and, though derived from observation, do not require more of it than has been made by children in general.

3. The demonstration is strictly logical, taking nothing for granted except the self-evident first principles, resting

nothing upon probability, and entirely independent of authority and opinion.

4. When the conclusion is attained by reasoning, its truth or falsehood can be ascertained, in geometry by actual measurement, in algebra by common arithmetical calculation. This gives confidence, and is absolutely necessary, if, as was said before, reason is not to be the instructor, but the pupil.

5. There are no words whose meanings are so much alike that the ideas which they stand for may be confounded. Between the meanings of terms there is no distinction, except a total distinction, and all adjectives and adverbs expressing difference of degrees are avoided. Thus it may be necessary to say, "A is greater than B;" but it is entirely unimportant whether A is very little or very much greater than B. Any proposition which includes the foregoing assertion will prove its conclusion generally, that is, for all cases in which A is greater than B, whether the difference be great or little. Locke mentions the distinctness of mathematical terms, and says in illustration: "The idea of two is as distinct from the idea of three as the magnitude of the whole earth is from that of a mite. This is not so in other simple modes, in which it is not so easy, nor perhaps possible for us to distinguish between two approaching ideas, which yet are really different; for who will undertake to find a difference between the white of this paper, and that of the next degree to it?"

These are the principal grounds on which, in our opinion, the utility of mathematical studies may be shown to rest, as a discipline for the reasoning powers. But the habits of mind which these studies have a tendency to form are valuable in the highest degree. The most important of all is the power of concentrating the ideas which a successful study of them increases where it did exist, and creates where it did not. A difficult position, or a new method of passing from one proposition to another, arrests all the attention and forces the united faculties to use their utmost exertions. The habit of mind thus formed soon extends itself to other pursuits, and is beneficially felt in all the business of life.

As a key to the attainment of other sciences, the use of the mathematics is too well known to make it necessary that we should dwell on this topic. In fact, there is not in this country any disposition to under-value them as regards the utility of their applications. But though they are now generally considered as a part, and a necessary one, of a liberal education, the views which are still taken of them as a part of education by a large proportion of the community are still very confined.

The elements of mathematics usually taught are contained in the sciences of arithmetic, algebra, geometry, and trigonometry. We have used these four divisions because they are generally adopted, though, in fact, algebra and geometry are the only two of them which are really distinct. Of these we shall commence with arithmetic, and take the others in succession in the order in which we have arranged them.

CHAPTER II.

ON ARITHMETICAL NOTATION.

THE first ideas of arithmetic, as well as those of other sciences, are derived from early observation. How they come into the mind it is unnecessary to inquire; nor is it possible to define what we mean by number and quantity. They are terms so simple, that is, the ideas which they stand for are so completely the first ideas of our mind, that it is impossible to find others more simple, by which we may explain them. This is what is meant by defining a term; and here we may say a few words on definitions in general, which will apply equally to all sciences.

Definition is the explaining a term by means of others, which are more easily understood, and thereby fixing its meaning, so that it may be distinctly seen what it does imply, as well as what it does not. Great care must be taken that the definition itself is not a tacit assumption of some fact or other which ought to be proved. Thus, when it is said that a square is "a four-sided figure, all whose sides are equal, and all whose angles are right angles," though no more is said than is true of a square, yet more is said than is necessary to define it, because it can be proved that if a four-sided figure have all its sides equal, and one only of its angles a right angle, all the other angles must be right angles also. Therefore, in making the above definition, we do, in fact, affirm that which ought to be proved. Again, the above definition, though redundant in one point, is, strictly speaking, defective in another, for it omits to state whether the sides of the figure are straight lines or curves. It should be, "a square is a four-sided rectilinear figure, all of whose sides are equal, and one of whose angles is a right angle."

As the mathematical sciences owe much, if not all, of the superiority of their demonstrations to the precision with which the terms are defined, it is most essential that the beginner should see clearly in what a good definition consists. We have seen that there are terms which cannot be defined, such as number and quantity. An attempt at a definition would only throw a difficulty in the student's way, which is already done in geometry by the attempts at an explanation of the terms point, straight line, and others, which are to be found in treatises on that subject. A point is defined to be that "which has no parts, and which has no magnitude;" a straight line is that which "lies evenly between its extreme points." Now, let any one ask himself whether he could have guessed what was meant, if, before he began geometry, any one had talked to him of "that which has no parts and which has no magnitude," and "the line which lies evenly between its extreme points," unless he had at the same time mentioned the words "point" and "straight line," which would have removed the difficulty? In this case the explanation is a great deal harder than the term to be explained, which must always happen whenever we are guilty of the absurdity of attempting to make the simplest ideas yet more simple.

A knowledge of our method of reckoning, and of writing

down numbers, is taught so early, that the method by which we began is hardly recollected. Few, therefore, reflect upon the very commencement of arithmetic, or upon the simplicity and elegance with which calculations are conducted. We find the method of reckoning by ten in our hands, we hardly know how, and we conclude, so natural and obvious does it seem, that it came with our language, and is a part of it; and that we are not much indebted to instruction for so simple a gift. It has been well observed, that if the whole earth spoke the same language, we should think that the name of any object was not a mere sign *chosen* to represent it, but was a sound which had some real connexion with the thing; and that we should laugh at, and perhaps persecute, any one who asserted that any other sound would do as well if we chose to think so. We cannot fall into this error, because, as it is, we happen to know that what we call by the sound "horse," the Romans distinguished as well by that of "*equus*," but we commit a similar mistake with regard to our system of numeration, because at present it happens to be received by all civilised nations, and we do not reflect on what was done formerly by almost all the world, and is done still by savages. The following considerations will, perhaps, put this matter on a right footing, and show that in our ideas of arithmetic we have not altogether rid ourselves of the tendency to attach ideas of mysticism to numbers which has prevailed so extensively in all times.

We know that we have nine signs to stand for the first nine numbers, and one for nothing, or zero. Also, that to represent ten we do not use a new sign, but combine two of the others, and denote it by 10, eleven by 11, and so on. But why was the number *ten* chosen as the limit of our separate symbols—why not nine, eight, or eleven? If we recollect how apt we are to count on the fingers, we shall be at no loss to see the reason. We can imagine our system of numeration formed thus:—A man proceeds to count a number, and to help the memory he puts a finger on the table for each one which he counts. He can thus go as far as ten, after which he must begin again, and by reckoning the fingers a second time he will have counted twenty, and so on. But this is not enough; he must also reckon the number of times which he has done this, and as by counting on the fingers he has divided the things which he is counting into lots of ten each, he may consider each lot as a unit of its kind, just as we say a number of sheep is *one* flock, twenty shillings are *one* pound. Call each lot a *ten.* In this way he can count a ten of tens, which he may call a hundred, a ten of hundreds, or a thousand, and so on. The process of reckoning would then be as follows:—Suppose, to choose an example, a number of faggots is to be counted. They are first tied up in bundles of ten each, until there are not so many as ten left. Suppose there are seven over. We then count the bundles of ten as we counted the single faggots, and tie them up also by tens, forming new bundles of one hundred each with some bundles of ten remaining. Let these last be six in number. We then tie up the bundles of hundreds by tens, making bundles of thousands, and find that there are five bundles of hundreds remaining. Suppose that on attempting to tie up the thousands by tens, we find there are not so many as ten, but only four. The number of faggots is then 4 thousands, 5 hundreds, 6 tens, and 7.

The next question is, how shall we represent this number in a short and convenient manner? It is plain that the way to do this is a *matter of choice*. Suppose then that we distinguish the tens by marking their number with one accent, the hundreds with two accents, and the thousands with three. We may then represent this number in any of the following ways:— $76'5''4'''$, $6'75''4'''$, $6'4'''5''7$, $4'''5''6'7$, the whole number of ways being 24. But this is more than we want; one certain method of representing a number is sufficient. The most natural way is to place them in order of magnitude, either putting the largest collection first or the smallest; thus $4'''5''6'7$, or $76'5''4'''$. Of these we choose the first.

In writing down numbers in this way it will soon be apparent that the accents are unnecessary. Since the singly accented figure will always be the second from the right, and so on, the *place* of each number will point out what accents to write over it, and we may therefore consider each figure as deriving a value from the place in which it stands. But here this difficulty occurs. How are we to represent the numbers $3'''3'$, and $4'''2'7$ without accents? If we write them thus, 33 and 427, they will be mistaken for $3'3$ and $4''2'7$. This difficulty will be obviated by placing cyphers so as to bring each number into the place allotted to the sort of collection which it represents; thus, since the trebly accented letters, or thousands, are in the fourth place from the right, and the singly accented letters in the second, the first number may be written 3030, and the second 4027. The cypher, which plays so important a part in arithmetic that it was anciently called the *art of cypher*, or *cyphering*, does not stand for any number in itself, but is merely employed, like blank types in printing, to keep other signs in those places which they must occupy in order to be read rightly. We may now ask what would have been the case if, instead of ten fingers, men had had more or less. For example, by what signs would 4567 have been represented, if man had nine fingers instead of ten? We may presume that the method would have been the same, with the number nine represented by 10 instead of ten, and the omission of the symbol 9. Suppose this number of faggots is to be counted by nines. Tie them up in bundles of nine, and we shall find 4 faggots remaining. Tie these bundles again in bundles of nine, each of which will, therefore, contain eighty-one, and there will be 3 bundles remaining. These tied up in the same way into bundles of nine, each of which contains seven hundred and twenty-nine, will leave 2 odd bundles, and, as there will be only six of them, the process cannot be carried any further. If, then, we represent, by $1'$, a bundle of nine, or a *nine*, by $1''$ a nine of nines, and so on, the number which we write 4567, must be written $6'''2''3'4$. In order to avoid confusion, we will suffer the accents to remain over all numbers which are not reckoned in tens, while those which are so reckoned shall be written in the common way. The following is a comparison of the way in which numbers in the common system are written, and in the one which we have just explained:

COUNTING BY

Tens	1	2	3	4	5	6	7	8	9	10	11	12	13
Nines	1	2	3	4	5	6	7	8	1′0	1′1	1′2	1′3	1′4

Tens	14	15	16	17	18	19	20	30	40	50
Nines	1′5	1′6	1′7	1′8	2′0	2′1	2′2	3′3	4′4	5′5

Tens	60	70	80	90	100
Nines	6′6	7′7	8′8	1″1′0	1″2′1

We will now write, in the common way, in the tens' system, the process which we went through in order to find how to represent the number 4567 in that of the nines, thus:

9) 4567
9) 507—rem. 4.
9)   56—rem. 3.
9)   6—rem. 2.
      0—rem. 6. Representation required, 6‴2″3′4.

The processes of arithmetic are the same in principle whatever system of numeration is used. To show this, we subjoin a question in each of the first four rules, worked both in the common system, and in that of the nines. There is the difference, that, in the first, the tens must be carried, and in the second the nines.

ADDITION.

$$\begin{array}{r} 636 \\ 987 \\ \underline{403} \\ 2026 \end{array} \qquad \begin{array}{r} 7''7'6 \\ 1'''3''1'6 \\ \underline{4''8'7} \\ 2'''7''0'1 \end{array}$$

SUBTRACTION.

$$\begin{array}{r} 1384 \\ \underline{797} \\ 587 \end{array} \qquad \begin{array}{r} 1'''8''0'7 \\ \underline{1'''0''7'5} \\ 7''2'2 \end{array}$$

MULTIPLICATION.

$$\begin{array}{r} 297 \\ \underline{136} \\ 1782 \\ 891 \\ \underline{297} \\ 40392 \end{array} \qquad \begin{array}{rrrrr} & & 3'' & 6' & 0 \\ & & \underline{1''} & \underline{6'} & \underline{1} \\ & & 3 & 6 & 0 \\ 2 & 4 & 0 & 0 & \\ \underline{3} & \underline{6} & \underline{0} & & \\ 6'''' & 1''' & 3'' & 6' & 0 \end{array}$$

DIVISION.

$$\begin{array}{l} 633)\ 79125\ (125 \\ \underline{633} \\ 1582 \\ \underline{1266} \\ 3165 \\ \underline{3165} \\ 0 \end{array} \qquad \begin{array}{lllllll} 7''7'3) & 1^{v} & 3^{iv} & 0^{iii} & 4^{ii} & 7^{i} & 6^{*}\ (1''4'8 \\ & \underline{7} & \underline{7} & \underline{3} & & & \\ & 4 & 2 & 1 & 7 & & \\ & \underline{3} & \underline{4} & \underline{2} & \underline{3} & & \\ & & 6 & 8 & 4 & 6 & \\ & & \underline{6} & \underline{8} & \underline{4} & \underline{6} & \\ & & & & & 0 & \end{array}$$

*To avoid too great a number of accents, Roman numerals are put instead of them; also, to avoid confusion, the accents are omitted after the first line.

The student should accustom himself to work questions in different systems of numeration, which will give him a clearer insight into the nature of arithmetical processes than he could obtain by any other method. When he uses a system in which numbers are counted by a number greater than ten, he will want some new symbols for figures. For example, in the duodecimal system, where twelve is the number of figures supposed, twelve will be represented by $1'0$; there must, therefore, be a distinct sign for ten and eleven: a nine and six reversed, thus ϱ and ∂, might be used for these.

CHAPTER III.

ELEMENTARY RULES OF ARITHMETIC.

As Soon as the beginner has mastered the notion of arithmetic, he may be made acquainted with the meaning of the algebraical signs $+$, $-$, $\times$, $=$, and also with that for division, or the common way of representing a fraction. There is no difficulty in these signs or in their use. Five minutes' consideration will make the symbol $5 + 3$ present as clear an idea as the words "5 added to 3." The reason why they usually cause so much embarrassment is, that they are generally deferred until the student commences algebra, when he is often introduced at the same time to the representation of numbers by letters, the distinction of known and unknown quantities, the signs of which we have been speaking, and the use of figures as the exponents of letters. Either of these four things is quite sufficient at a time, and there is no time more favorable for beginning to make use of the signs of operation than when the habit of performing the operations commences. The beginner should exercise himself in putting the simplest truths of arithmetic in this new shape, and should write such sentences as the following frequently:

$$2 + 7 = 9,$$

$$6 - 4 = 2,$$

$$1 + 8 + 4 - 6 = 4 + 2 + 1,$$

$$2 \times 2 + 12 \times 12 = 14 \times 10 + 2 \times 2 \times 2.$$

These will accustom him to the meaning of the signs, just as he was accustomed to the formation of letters by writing copies. As he proceeds through the rules of arithmetic, he should take care never to omit connecting each operation with its sign, and should avoid confounding operations together and considering them as the same, because they produce the same result. Thus 4×7 does not denote the same operation as 7×4, though the result of both is 28. The first is four multiplied by seven, four taken seven times; the second is seven multiplied by four, seven taken four times; and that $4 \times 7 = 7 \times 4$ is a proposition to be proved, not to be taken for granted. Again, $\frac{1}{7} \times 4$ and $\frac{4}{7}$ are marks of distinct operations, though their result is the same, as we shall show in treating of fractions.

The examples which a beginner should choose for practice should be simple and should not contain very large numbers. The powers of the mind cannot be directed to two things at once: if the complexity of the numbers used requires all the student's attention, he cannot observe the principle of the rule which he is following. Now, at the commencement of his career, a principle is not received and understood by the student as quickly as it is explained by the instructor. He does not, and cannot, generalise at all; he must be taught to do so; and he cannot learn that a particular fact holds good for *all numbers* unless by having it shown that it holds good for *some numbers*, and that for those *some numbers* he may substitute *others*, and use the same demonstration. Until he can do this himself he does not understand the principle, and he can never do this except by seeing the rule explained and trying it himself on small numbers. He may, indeed, and will, believe it on the word of his instructor, but this disposition is to be checked. He must be told, that whatever is not gained by his own thought is not gained to any purpose; that the mathematics are put in his way purposely because they are the only sciences in which he must not trust the authority of any one. The superintendence of these efforts is the real business of an instructor in arithmetic. The merely showing the student a rule by which he is to work, and comparing his answer with a key to the book, printed for the preceptor's private use, to save the trouble which he ought to bestow upon his pupil, is not teaching arithmetic any more than presenting him with a grammar and dictionary is teaching him Latin. When the principle of each rule has been well established by showing its application to some simple examples (and the number of these requisite will vary with the intellect of the student), he may then proceed to more complicated cases, in order to acquire facility in computation. The four first rules may be studied in this way, and these will throw the greatest light on those which succeed.

The student must observe that all operations in arithmetic may be resolved into addition and subtraction; that these additions and subtractions might be made with counters; so that the whole of the rules consist of processes intended to shorten and simplify that which would otherwise be long and complex. For example, multiplication is continued addition of the same number to itself—twelve times seven is twelve sevens added together. Division is a continued subtraction of one number from another; the division of 129 by 3 is a continued subtraction of 3 from 129, in order to see how many threes it contains. All other operations are composed of these four, and are, therefore, the result of additions and subtractions only.

The following principles, which occur so continually in mathematical operations that we are, at length, hardly sensible of their presence, are the foundation of the arithmetical rules:

I. We do not alter the sum of two numbers by taking away any part of the first, if we annex that part to the second. This may be expressed by signs, in a particular instance, thus:

$$(20 - 6) + (32 + 6) = 20 + 32.$$

II. We do not alter the difference of two numbers by increasing or diminishing one of them, provided we increase or diminish the other as much. This may be expressed thus, in one instance:

$$(45 + 7) - (22 + 7) = 45 - 22,$$
$$(45 - 8) - (22 - 8) = 45 - 22.$$

III. If we wish to multiply one number by another, for example 156 by 29, we may break up 156 into any number of parts, multiply each of these parts by 29, and add the results. For example, 156 is made up of 100, 50, and 6. Then

$$156 \times 29 = 100 \times 29 + 50 \times 29 + 6 \times 29.$$

IV. The same thing may be done with the multiplier instead of the multiplicand. Thus, 29 is made up of 18, 6, and 5. Then

$$156 \times 29 = 156 \times 18 + 156 \times 6 + 156 \times 5.$$

V. If any two or more numbers be multiplied together, it is indifferent in what order they are multiplied, the result is the same. Thus,

$$10 \times 6 \times 4 \times 3 = 3 \times 10 \times 4 \times 6 = 6 \times 10 \times 4 \times 3, \quad \text{etc.}$$

VI. In dividing one number by another, for example 156 by 12, we may break up the dividend, and divide each of its parts by the divisor, and then add the results. We may part 156 into 72, 60, and 24; this is expressed thus:

$$\frac{156}{12} = \frac{72}{12} + \frac{60}{12} + \frac{24}{12}.$$

The same thing cannot be done with the divisor. It is not true that

$$\frac{156}{12} = \frac{156}{3} + \frac{156}{4} + \frac{156}{5}.$$

The student should discover the reason for himself.

A prime number is one which is not divisible by any other number except 1. When the process of division can be performed, it can be ascertained whether a given number is divisible by any other number, that is, whether it is prime or not. This can be done by dividing it by all the numbers which are less than its half, since it is evident that it cannot be divided into a number of parts, each of which is greater than its half. This process would be laborious when the given number is large; still it may be done, and by this means the number itself may be *reduced to its prime factors*,* as it is called, that is, it may either be shown to be a prime number itself or made up by multiplying several prime numbers together. Thus, 306 is 34×9, or $2 \times 17 \times 9$, or $2 \times 17 \times 3 \times 3$, and has for its prime factors 2, 17, and 3, the latter of which is repeated twice in its formation. When this has been done with two numbers, we can then see whether they have any factors in common, and, if that be the case, we can then find what is called their *greatest common measure* or *divisor*; that is, the number made by multiplying all their common factors. It is an evident truth that, if a number can be divided by the product of two others, it can be divided by each of them. If a number can be parted into an exact number of twelves, it can be parted also into a number of sixes, twos, or fours. It is also true that, if a number can be divided by any other number, and the quotient can then be divided by a third number, the original number can be divided by the product of the other two. Thus, 144 is divisible by 2; the quotient, 72, is divisible by 6; and the original number is divisible by 6×2 or 12. It is also true that, if two numbers are prime, their product is divisible by no numbers except themselves. Thus,

*The *factors* of a number are those numbers by the multiplication of which it is made.

17×11 is divisible by no numbers except 17 and 11. Though this is a simple proposition, its proof is not so, and cannot be given to the beginner. From these things it follows that the greatest common measure of two numbers (measure being an old word for divisor) is the product of all the prime factors which the two possess in common. For example, the numbers 90 and 100, which, when reduced to their prime factors, are $2 \times 5 \times 3 \times 3$ and $2 \times 2 \times 5 \times 5$, have the common factors 2 and 5, and are divisible by 2×5, or 10. The quotients are 3×3 and 2×5, or 9 and 10, which have no common factor remaining, and 2×5, or 10, is the greatest common measure of 90 and 100. The same may be shown in the case of any other numbers. But the method we have mentioned of resolving numbers into their prime factors, being troublesome to apply when the numbers are large, is usually abandoned for another. It happens frequently that a method simple in principle is laborious in practice, and the contrary.

When one number is divided by another, and its quotient and remainder obtained, the dividend may be recovered again by multiplying the quotient and divisor together, and adding the remainder to the product. Thus 171 divided by 27 gives a quotient 6 and a remainder 9, and 171 is made by multiplying 27 by 6, and adding 9 to the product. That is, $171 = 27 \times 6 + 9$. Now, from this equation it is easy to show that every number which divides 171 and 27 also divides 9, that is, every common measure of 171 and 27 is also a common measure of 27 and 9. We can also show that 27 and 9 have no common measures which are not common to 171 and 27. Therefore, the common measures of 171 and 27 are those, and no others, which are common to 27 and 9; the greatest common measure of each pair must, therefore, be the same, that is, the greatest common measure of a divisor and dividend is also the greatest common measure of the remainder and divisor. Now take the common process for finding the greatest common measure of two numbers; for example, 360 and 420, which is as follows, and abbreviate the words *greatest common measure* into their initials *g. c. m.*:

```
360)420(1
    360
    ---
     60)360(6
        360
        ---
          0
```

From the theorem above enunciated it appears that

g. c. m. of 420 and 360 is *g. c. m.* of 60 and 360;
g. c. m. of 60 and 360 is 60;

because 60 divides both 60 and 360, and no number can have a greater measure than itself. Thus may be seen the reason of the common rule for finding the greatest common measure of two numbers.

Every number which can be divided by another without remainder is called a multiple of it. Thus, 12, 18, and 42 are multiples of 6, and the last is a *common multiple* of 6 and 7, because it is divisible both by 6 and 7. The only things which it is necessary to observe on this subject are, (1), that the product of two numbers is a common multiple of both; (2), that when the two numbers have a common measure greater than 1, there is a common multiple less than their product; (3), that when they have no common measure except 1, the least common multiple is their product. The first of these is evident; the second will appear from an example. Take 10 and 8, which have the common measure 2, since the first is 2×5 and the second 2×4. The product is $2\times2\times4\times5$, but $2 \times 4 \times 5$ is also a common multiple, since it is divisible by 2×4, or 8, and by 2×5, or 10. To find this common multiple we must, therefore, divide the product by the greatest common measure. The third principle cannot be proved in an elementary way, but the student may convince himself of it by any number of examples. He will not, for instance, be able to find a common multiple of 8 and 7 less than 8×7 or 56.

CHAPTER IV.

ARITHMETICAL FRACTIONS.

WHEN the student has perfected himself in the four rules, together with that for finding the greatest common measure, he should proceed at once to the subject of fractions. This part of arithmetic is usually supposed to present extraordinary difficulties; whereas, the fact is that there is nothing in fractions so difficult, either in principle or practice, as the rule for finding the greatest common measure. We would recommend the student not to attend to the distinctions of proper and improper, pure or mixed fractions, etc., as there is no distinction whatever in the rules, which are common to all these fractions.

When one number, as 56, is to be divided by another, as 8, the process is written thus: $\frac{56}{8}$. By this we mean that 56 is to be divided into 8 equal parts, and one of these parts is called the quotient. In this case the quotient is 7. But it is equally possible to divide 57 into 8 equal parts; for example, we can divide 57 feet into 8 equal parts, but the eighth part of 57 feet will not be an exact number of feet, since 57 does not contain an exact number of eights; a part of a foot will be contained in the quotient $\frac{57}{8}$, and this quotient is therefore called a fraction, or broken number. If we divide 57 into 56 and 1, and take the eighth part of each of these, whose sum will give the eighth part of the whole, the eighth of 56 feet is 7 feet; the eighth of 1 foot is a fraction, which we write $\frac{1}{8}$, and $\frac{57}{8}$ is $7 + \frac{1}{8}$, which is usually written $7\frac{1}{8}$. Both of these quantities $\frac{57}{8}$, and $7\frac{1}{8}$, are called fractions; the only difference is that, in the second, that part of the quotient which is a whole number is separated from the part which is less than any whole number.

There are two ways in which a fraction may be considered. Let us take, for example, $\frac{5}{8}$. This means that 5 is to be divided into 8 parts, and $\frac{5}{8}$ stands for one of these parts. The same length will be obtained if we divide 1 into 8 parts, and take 5 of them, or find $\frac{1}{8} \times 5$. To prove this let each of the lines drawn below represent $\frac{1}{8}$ of an inch; repeat $\frac{1}{8}$ five times, and repeat the same line eight times.

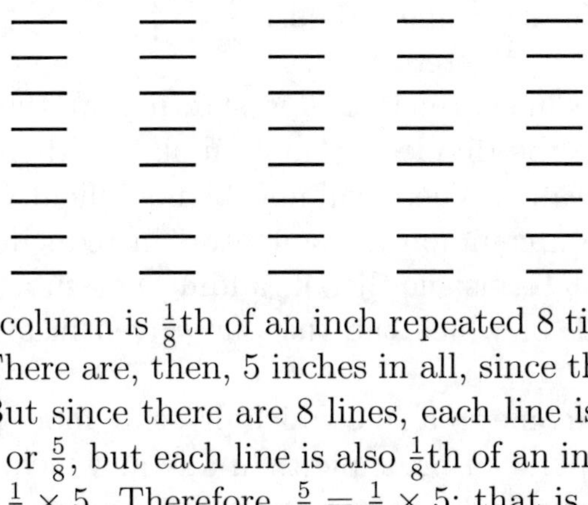

In each column is $\frac{1}{8}$th of an inch repeated 8 times; that is one inch. There are, then, 5 inches in all, since there are five columns. But since there are 8 lines, each line is the eighth of 5 inches, or $\frac{5}{8}$, but each line is also $\frac{1}{8}$th of an inch repeated 5 times, or $\frac{1}{8} \times 5$. Therefore, $\frac{5}{8} = \frac{1}{8} \times 5$; that is, in order to find $\frac{5}{8}$ inches, we may either divide *five inches* into 8 parts, and take *one* of them, or divide *one inch* into 8 parts, and take *five* of them. The symbol $\frac{5}{8}$ is made to stand for both these operations, since they lead to the same result.

The most important property of a fraction is, that if both its numerator and denominator are multiplied by the same number, the value of the fraction is not altered; that is, $\frac{3}{5}$ is the same as $\frac{12}{20}$, or each part is the same when we divide 12 inches into 20 parts, as when we divide 3 inches into 5 parts. Again, we get the same length by dividing 1 inch into 20 parts, and taking 12 of them, which we get by dividing 1 inch into 5 parts and taking 3 of them. This hardly needs demonstration. Taking 12 out of 20 is taking 3 out of 5, since for every 3 which 12 contains, there is a 5 contained in 20. Every fraction, therefore, admits of innumerable alterations in its form, without any alteration in its value. Thus, $\frac{1}{2} = \frac{2}{4} = \frac{3}{6} = \frac{4}{8} = \frac{5}{10}$, etc.; $\frac{2}{7} = \frac{4}{14} = \frac{6}{21} = \frac{8}{28}$, etc.

On the same principle it is shown that the terms of a fraction may be divided by any number without any alteration of its value. There will now be no difficulty in reducing fractions to a common denominator, in reducing a fraction to its lowest terms; neither in adding nor subtracting fractions, for all of which the rules are given in every book of arithmetic.

We now come to a rule which presents more peculiar difficulties in point of principle than any at which we have yet arrived. If we could at once take the most general view of numbers, and give the beginner the extended notions which he may afterwards attain, the mathematics would present comparatively few impediments. But the constitution of our minds will not permit this. It is by collecting facts and principles, one by one, and thus only, that we arrive at what are called general notions; and we afterwards make comparisons of the facts which we have acquired and discover analogies and resemblances which, while they bind together the fabric of our knowledge, point out methods of increasing its extent and beauty. In the limited view which we first take of the operations which we are performing, the names which we give are necessarily confined and partial; but when, after additional study and reflection, we recur to our former notions, we soon discover processes so resembling one another, and different rules so linked together, that we feel it would destroy the symmetry of our language if we were to call them by different names. We are then induced to extend the meaning of our terms, so as to make two rules into one. Also, suppose that when we have discovered and applied a rule and given the process which it teaches a particular name, we find that this process is only a part of one more general, which applies to all cases contained under the first, and to others besides. We have only the alternative of inventing a new name, or of extending the meaning of the former one so as to merge the particular process in the more general one of which it is a part. Of this we can give an instance. We began with reasoning upon simple numbers, such as 1, 2, 3, 20, etc. We afterwards divided these into parts, of which we took some number, and which we called fractions, such as $\frac{2}{3}$, $\frac{7}{2}$, $\frac{1}{5}$, etc. Now there is no number which may not be considered as a fraction in as many different ways as we please. Thus 7 is $\frac{14}{2}$ or $\frac{21}{3}$, etc.; 12 is $\frac{144}{12}$, $\frac{72}{6}$, etc. Our new notion of fraction is, then, one which includes all our former ideas of number, and others besides. It is then customary to represent by the word number, not only our first notion of it, but also the extended one, of which the first is only a part. Those to which our first notions applied we call whole numbers, the others fractional numbers, but still the name number is applied to both 2 and $\frac{1}{2}$, 3 and $\frac{3}{5}$. The rule of which we have spoken is another instance. It is called the multiplication of fractional numbers. Now, if we return to our meaning of the word multiplication, we shall find that the multiplication of one fraction by another appears an absurdity. We multiply a number by taking it several times and adding these together. What, then, is meant by multiplying by a fraction? Still, a rule has been found which, in applying mathematics, it is necessary to use for fractions, in all cases where multiplication would have been used had they been

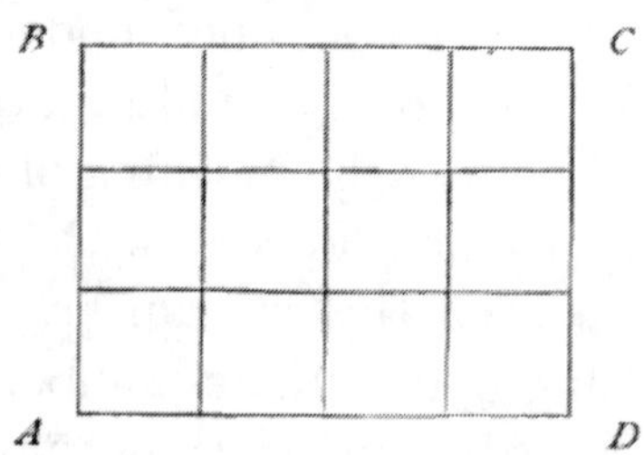

whole numbers. Of this we shall now give a simple example. Take an oblong figure (which is called a rectangle in geometry), such as $ABCD$, and find the magnitudes of the sides

AB and BC in inches. Draw the line EF equal in length to one inch, and the square G, each of whose sides is one inch. If the lines AB and BC contain an exact number of inches, the rectangle $ABCD$ contains an exact number of squares, each

equal to G, and the number of squares contained is found by multiplying the number of inches in AB by the number of inches in BC. In the present case the number of squares is 3×4, or 12. Now, suppose another rectangle $A'B'C'D'$, of which neither of the sides is an exact number of inches; suppose, for example, that $A'B'$ is $\frac{2}{3}$ of an inch, and that $B'C'$ is $\frac{5}{7}$ of an inch. We may show, by reasoning, that we can find

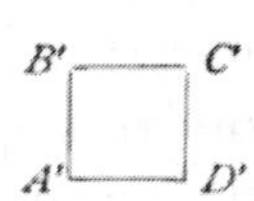

how much $A'B'C'D'$ is of G by forming a fraction which has the product of the numerators of $\frac{2}{3}$ and $\frac{5}{7}$ for its numerator, and the product of their denominators for its denominator; that is, that $A'B'C'D'$ contains $\frac{10}{21}$ of G. Here then appears a connexion between the multiplication of whole numbers, and the formation of a fraction, whose numerator is the product of two numerators, and its denominator the product of the corresponding denominators. These operations will always come together, that is whenever a question occurs in which, when whole numbers are given, those numbers are to be multiplied together; when fractional numbers are given, it will be necessary, in the same case, to multiply the numerator by the numerator, and the denominator by the denominator, and form the result into a fraction, as above.

This would lead us to suspect some connexion between these two operations, and we shall accordingly find that when whole numbers are formed into fractions, they may be multiplied together by this very rule. Take, for example, the numbers 3 and 4, whose product is 12. The first may be written as $\frac{15}{5}$, and the second as $\frac{8}{2}$. Form a fraction from the product of the numerators and denominators of these, which will be $\frac{120}{10}$, which is 12, the product of 3 and 4.

From these considerations it is customary to call the fraction which is produced from two others in the manner above stated, the *product* of those two fractions, and the process of finding the third fraction, *multiplication.* We shall always find the first meaning of the word multiplication included in the second, in all cases in which the quantities represented as fractions are really whole numbers. The mathematics are not the only branches of knowledge in which it is customary to extend the meaning of established terms. Whenever we pass from that which is simple to that which is complex, we shall see the necessity of carrying our terms with us and enlarging their meaning, as we enlarge our own ideas. This is the only method of forming a language which shall approach in any degree towards perfection; and more depends upon a well-constructed language in mathematics than in anything else. It is not that an imperfect language would deprive us of the means of demonstration, or cramp the powers of reasoning. The propositions of Euclid upon numbers are as rationally established as any others, although his terms are deficient in analogy, and his notation infinitely inferior to that which we use. It is the progress of discovery which is checked by terms constructed so as to conceal resemblances which exist, and to prevent one result from pointing out another. The higher branches of mathematics date the progress which they have made in the last century and a half, from the time when the genius of Newton, Leibnitz, Descartes, and Hariot turned the attention of the scientific world to the imperfect mechanism of the science. A slight and almost casual improvement, made by Hariot in algebraical language, has been the foundation of most important branches of the science.* The subject of the last articles is of very great importance, and will often recur to us in explaining the difficulties of algebraical notation.

The multiplication of $\frac{5}{6}$ by $\frac{3}{2}$ is equivalent to dividing $\frac{5}{6}$ into 2 parts, and taking three such parts. Because $\frac{5}{6}$ being the same as $\frac{10}{12}$, or 1 divided into 12 parts and 10 of them taken, the half of $\frac{10}{12}$ is 5 of those parts, or $\frac{5}{12}$. Three times this quantity will be 15 of those parts, or $\frac{15}{12}$, which is by our rule the same as what we have called, $\frac{5}{6}$ multiplied by $\frac{3}{2}$. But

*The mathematician will be aware that I allude to writing an equation in the form $x^2 + ax - b = 0$; instead of $x^2 + ax = b$.

the same result arises from multiplying $\frac{3}{2}$ by $\frac{5}{6}$, or dividing $\frac{3}{2}$ into 6 parts and taking 5 of them. Therefore, we find that $\frac{3}{2}$ multiplied by $\frac{5}{6}$ is the same as $\frac{5}{6}$ multiplied by $\frac{3}{2}$, or $\frac{3}{2} \times \frac{5}{6} = \frac{5}{6} \times \frac{3}{2}$. This proposition is usually considered as requiring no proof, because it is received very early on the authority of a rule in the elements of arithmetic. But it is not self-evident, for the truth of which we appeal to the beginner himself, and ask him whether he would have seen at once that $\frac{5}{6}$ of an apple divided into 2 parts and 3 of them taken, is the same as $\frac{3}{2}$ of an apple, or one apple and a-half divided into six parts and 5 of them taken.

An extension of the same sort is made of the term division. In dividing one whole number by another, for example, 12 by 2, we endeavor to find how many *twos* must be added together to make 12. In passing from a problem which contains these whole numbers to one which contains fractional quantities, for example $\frac{3}{4}$ and $\frac{2}{5}$, it will be observed that in place of finding how many twos make 12, we shall have to find into how many parts $\frac{2}{5}$ must be divided, and how many of them must be taken, so as to give $\frac{3}{4}$. If we reduce these fractions to a common denominator, in which case they will be $\frac{15}{20}$ and $\frac{8}{20}$; and if we divide the second into 8 equal parts, each of which will be $\frac{1}{20}$, and take 15 of these parts, we shall get $\frac{15}{20}$, or $\frac{3}{4}$. The fraction whose numerator is 15, and whose denominator is 8, or $\frac{15}{8}$, will in these problems take the place of the quotient of the two whole numbers. In the same manner as before, it may be shown that this process is equivalent to the division of one whole number by another, whenever the fractions are really whole numbers; for example, 3 is $\frac{12}{4}$, and 15 is $\frac{30}{2}$. If this process be applied to $\frac{30}{2}$ and $\frac{12}{4}$, the result is $\frac{120}{24}$, which is 5, or the same as 15 divided by 3. This process is then, by extension, called division: $\frac{15}{8}$ is called the quotient of $\frac{3}{4}$ divided by $\frac{2}{5}$, and is found by multiplying the numerator of the first by the denominator of the second for the numerator of the result, and the denominator of the first by the numerator of the second for the denominator of the result. That this process does give the same result as ordinary division in all cases where ordinary division is applicable, we can easily show from any two whole numbers, for example, 12 and 2, whose quotient is 6. Now 12 is $\frac{36}{3}$, and 2 is $\frac{10}{5}$, and the rule for what we have called division of fractions will give as the quotient $\frac{180}{30}$, which is 6.

In all fractional investigations, when the beginner meets with a difficulty, he should accustom himself to leave the notation of fractions, and betake himself to their original definition. He should recollect that $\frac{5}{6}$ is 1 divided into 6 parts and five of them taken, or the sixth part of 5, and he should reason upon these suppositions, neglecting all rules until he has established them in his own mind by reflexion on particular instances. These instances should not contain large numbers, and it will perhaps assist him if he reasons on some given unit, for example a foot. Let AB be one foot, and divide it into any number of equal parts (7 for example) by the points C, D, E, F, G, and H. He must then recollect that each of these parts is $\frac{1}{7}$ of a foot; that any two of them together are $\frac{2}{7}$ of a foot; any 3, $\frac{3}{7}$, and so on. He should

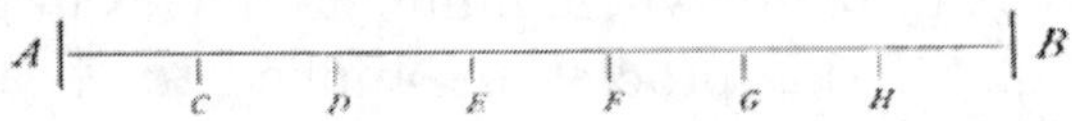

then accustom himself, without a rule, to solve such questions as the following, by observation of the figure, dividing each part into several equal parts, if necessary; and he may be well assured that he does not understand the nature of fractions until such questions are easy to him.

What is $\frac{1}{4}$ of $\frac{2}{7}$ of a foot? What is $\frac{2}{5}$ of $\frac{1}{3}$ of $\frac{3}{4}$ of a foot? Into how many parts must $\frac{3}{7}$ of a foot be divided, and how many of them must be taken to produce $\frac{14}{15}$ of a foot? What is $\frac{1}{3} + \frac{1}{7}$ of a foot? and so on.

CHAPTER V.

DECIMAL FRACTIONS.

IT is a disadvantage attending rules received without a knowledge of principles, that a mere difference of language is enough to create a notion in the mind of a student that he is upon a totally different subject. Very few beginners see that in following the rule usually called practice, they are working the same questions as were proposed in compound multiplication;—that the rule of three is only an application of the doctrine of fractions; that the rules known by the name of commission, brokerage, interest, etc., are the same, and so on. No instance, however, is more conspicuous than that of decimal fractions, which are made to form a branch of arithmetic as distinct from ordinary or vulgar fractions as any two parts of the subject whatever. Nevertheless, there is no single rule in the one which is not substantially the same as the rule corresponding in the other, the difference consisting altogether in a different way of writing the fractions. The beginner will observe that throughout the subject it is continually necessary to reduce fractions to a common denominator: he will see, therefore, the advantage of always using either the same denominator, or a set of denominators, so closely connected as to be very easily reducible to one another. Now of all numbers which can be chosen the most easily manageable are 10, 100, 1000, etc., which are called decimal numbers on account of their connexion with the number ten. All fractions, such as $\frac{75}{100}$, $\frac{333}{1000}$, $\frac{178699}{10}$, which have a decimal number for the denominator, are called decimal fractions. Now a denominator of this sort is known whenever the number of cyphers in it are known; thus a decimal number with 4 cyphers can only be 10,000, or ten thousand. We need not, therefore, write the denominator, provided, in its stead, we put some mark upon the numerator, by which we may know the number of cyphers in the denominator. This mark is for our own selection. The method which is followed is to point off from the numerator as many *figures* as there are *cyphers* in the denominator. Thus $\frac{17334}{1000}$ is represented by 17.334; $\frac{229}{1000}$ thus, .229. We might, had we so pleased, have represented them thus, 17334^3, 229^3; or thus, 17334_3, 229_3, or in any way by which we might choose to agree to recollect that the denominator is 1 followed by 3 cyphers. In the common method this difficulty occurs immediately. What shall be done when there are not as many figures in the numerator as there are cyphers in the denominator? How shall we represent $\frac{88}{10000}$? We must here extend our language a little, and imagine some method by which, without essentially altering the numerator, it may be made to show the number of cyphers in the denominator. Something of the sort has already been done in representing a number of tens, hundreds, or thousands, etc.; for 5 thousands were represented by 5000, in which, by the assistance of cyphers, the 5 is made to stand in the place allotted to thousands. If, in the present instance, we place cyphers at the beginning of the numerator, until the number of figures and cyphers together is equal to the number of cyphers in the denominator, and place a point before the first cypher, the fraction $\frac{88}{10000}$ will be represented thus, .0088; by which we understand a fraction whose numerator is 88, and whose denominator is a decimal number containing four cyphers.

There is a close connexion between the manner of representing decimal fractions, and the decimal notation for numbers. Take, for example, the fraction 217.3426 or $\frac{2173426}{10000}$. You will recollect that 2173426 is made up of $2000000 + 100000 + 70000 + 3000 + 400 + 20 + 6$. If each of these parts be divided by 10000, and the quotient obtained or the fraction reduced to its lowest terms, the result is as follows:

$$\frac{2173426}{10000} = 200 + 10 + 7 + \frac{3}{10} + \frac{4}{100} + \frac{2}{1000} + \frac{6}{10000}.$$

We see, then, that in the fraction 217.3426 the first figure 2 counts two hundred; the second figure, 1, ten, and the third 7 units. It appears, then, that all figures on the left of the decimal point are reckoned as ordinary numbers. But on the right of that point we find the figure 3, which counts for $\frac{3}{10}$; 4, which counts for $\frac{4}{100}$; 2, for $\frac{2}{1000}$; and 6, for $\frac{6}{10000}$. It appears therefore, that numbers on the right of the decimal point decrease as they move towards the right, each number being one-tenth of what it would have been had it come one place nearer to the decimal point. The first figure on the right hand of that point is so many tenths of a unit, the second figure so many hundredths of a unit, and so on.

The learner should go through the same investigation with other fractions, and should demonstrate by means of the principles of fractions, generally, such exercises as the following, until he is thoroughly accustomed to this new method of writing fractions:

$$.68342 = .6 + .08 + .003 + .0004 + .00002,$$

$$\text{or}\quad \frac{68342}{100000} = \frac{6}{10} + \frac{8}{100} + \frac{3}{1000} + \frac{4}{10000} + \frac{2}{100000},$$

$$.00012 = .0001 + .00002 = \frac{1}{10000} + \frac{2}{100000},$$

$$163.499 = \frac{163429}{1000} = 163\frac{429}{1000}$$

$$= \frac{1634}{10} + \frac{29}{1000} = \frac{16342}{100} + \frac{9}{1000}, \quad \text{etc.}$$

The rules of addition, subtraction, and multiplication may now be understood. In addition and subtraction, the keeping the decimal points under one another is equivalent to reducing the fractions to a common denominator, as we may show thus: Take two fractions, 1.5 and 2.125, or $\frac{15}{10}$ and $\frac{2125}{1000}$, which, reducing the first to the denominator of the second, may be written $\frac{1500}{1000}$ and $\frac{2125}{1000}$. If we add the numerators together, we find the sum of the fractions $\frac{3625}{1000}$, or 3.625

$$\begin{array}{r} 2125 \\ 1500 \\ \hline 3625 \end{array} \qquad \begin{array}{l} 2.125 \\ 1.5 \\ \hline 3.625 \end{array}$$

The learner can now see the connexion of the rule given

for the addition of decimal fractions with that for the addition of vulgar fractions. There is the same connexion between the rules of subtraction. The principle of the rule of multiplication is as follows: If two decimal numbers be multiplied together, the product has as many cyphers as are in both together. Thus $100 \times 1000 = 100000$, $10 \times 100 = 1000$, etc. Therefore the denominator of the product, which is the product of the denominators, has as many cyphers as are in the denominators of both fractions, and since the numerator of the product is the product of the numerators, the point must be placed in that product so as to cut off as many decimal places as are both in the multiplier and the multiplicand. Thus:

$$\frac{13}{100} \times \frac{12}{10} = \frac{156}{1000}, \quad \text{or} \quad .13 \times 1.2 = .156;$$

$$\frac{4}{1000} \times \frac{6}{100} = \frac{24}{100000},$$

$$\text{or} \quad .004 \times .06 = .00024, \quad \text{etc.}$$

It is a general rule, that wherever the number of figures falls short of what we know ought to be the number of decimals, the deficiency is made up by cyphers.

It may now be asked, whether all fractions can be reduced to decimal fractions? It may be answered that they cannot. It is a principle which is demonstrated in the science of algebra,—that if a number be not divisible by a prime number, no multiplication of that number, by itself, will make it so. Thus 10 not being divisible by 7, neither 10×10, nor

$10 \times 10 \times 10$, etc., is divisible by 7. A consequence of this is, that since 5 and 2 are the only prime numbers which will divide 10, no fraction can be converted into a decimal unless its denominator is made up of products, either of 5 or 2, or of both combined, such as 5×2, $5 \times 5 \times 2$, $5 \times 5 \times 5$, 2×2, etc. To show that this is the case, take any fraction with such a denominator; for example, $\frac{13}{5 \times 5 \times 5}$. Multiply the numerator and denominator by 2, once for every 5, which is contained in the denominator, and the fraction will then become

$$\frac{13 \times 2 \times 2 \times 2}{5 \times 5 \times 5 \times 2 \times 2 \times 2}, \quad \text{or} \quad \frac{2 \times 2 \times 2 \times 13}{10 \times 10 \times 10},$$

which is $\frac{140}{1000}$, or .104. In a similar way, any fraction whose denominator has no other factors than 2 or 5, can be reduced to a decimal fraction. We first search for such a number as will, when multiplied by the denominator, produce a decimal number, and then multiply both the numerator and denominator by that number.

No fraction which has any other factor in its denominator can be reduced to a decimal fraction exactly. But here it must be observed that in most parts of mathematical computation a very small error is not material. In different species of calculations, more or less exactness may be required; but even in the most delicate operations, there is always a limit beyond which accuracy is useless, because it cannot be appreciated. For example, in measuring land for sale, an error of an inch in five hundred yards is not worth avoiding, since even if such an error were committed, it would not make a difference which would be considered as of any consequence, as in all probability the expense of a more accurate measurement would be more than the small quantity of land thereby saved would be worth. But in the measurement of a line for the commencement of a trigonometrical survey, an error of an inch in five hundred yards would be fatal, because the subsequent processes involve calculations of such a nature that this error would be multiplied, and cause a considerable error in the final result. Still, even in this case, it would be useless to endeavor to avoid an error of one-thousandth part of an inch in five hundred yards; first, because no instruments hitherto made would show such an error: and secondly, because if they could, no material difference would be made in the result by a correction of it. Again, we know that almost all bodies are lengthened in all directions by heat. For example: A brass ruler which is a foot in length to-day, while it is cold, will be more than a foot to-morrow if it is warm. The difference, nevertheless, is scarcely, if at all, perceptible to the naked eye, and it would be absurd for a carpenter, in measuring a few feet of mahogany for a table, to attempt to take notice of it; but in the measurement of the base of a survey, which is several miles in length and takes many days to perform, it is necessary to take this variation into account, as a want of attention to it may produce perceptible errors in the result: nevertheless, any error which has not this effect, it would be useless to avoid even were it possible. We see, therefore, that we may, instead of a fraction, which cannot be reduced to a decimal, substitute a decimal fraction, if we can find one so near to the former, that the error committed by the substitution will not materially affect the result. We will now proceed to show how to find a series of decimal fractions, which approach nearer and nearer to a given fraction, and also that, in this approximation, we may approach as near as we please to the given fraction without ever being exactly able to reach it.

Take, for example,, the fraction $\frac{7}{11}$. If we divide the series of numbers 70, 700, 7000, etc., by 11, we shall obtain the following results:

$\frac{70}{11}$	gives the quotient	6, and the remainder	4, and is	$6\frac{4}{11}$,
$\frac{700}{11}$	"	63 "	7	$63\frac{7}{11}$,
$\frac{7000}{11}$	"	636 "	4	$636\frac{4}{11}$,
$\frac{70000}{11}$	"	6363 "	7	$6363\frac{7}{11}$,
etc.,		etc.,	etc.	

Now observe that if two numbers do not differ by so much as 1, their tenth parts do not differ by so much as $\frac{1}{10}$, their hundredth parts by so much as $\frac{1}{100}$, their thousandth parts by so much as $\frac{1}{1000}$, and so on; and also remember that $\frac{7}{11}$ is the tenth part of $\frac{70}{11}$, the hundredth part of $\frac{700}{11}$, and so on.

The two following tables will now be apparent:

$\frac{70}{11}$	does not differ from	6	by so much as	1,
$\frac{700}{11}$	"	63	"	1,
$\frac{7000}{11}$	"	636	"	1,
$\frac{70000}{11}$	"	6363	"	1,
etc.,		etc.,		etc.

Therefore

$\frac{7}{11}$	does not differ from	$\frac{6}{10}$	or	.6,	by so much as	$\frac{1}{10}$	or	.1,
$\frac{7}{11}$	"	$\frac{63}{100}$	"	.63	"	$\frac{1}{100}$	"	.01,
$\frac{7}{11}$	"	$\frac{636}{1000}$	"	.636	"	$\frac{1}{1000}$	"	.001,
$\frac{7}{11}$	"	$\frac{6363}{10000}$	"	.6363	"	$\frac{1}{10000}$	"	.0001,
etc.,		etc.,				etc.		

We have then a series of decimal fractions, viz., .6, .63, .636, .6363, .63636, etc., which continually approach more and more near to $\frac{7}{11}$, and therefore in any calculation in which the fraction $\frac{7}{11}$ appears, any one of these may be substituted for it, which is sufficiently near to suit the purpose for which the calculation is intended. For some purposes .636 would be a sufficient approximation; for others .63636363 would be necessary. Nothing but practice can show how far the approximation should be carried in each case.

The division of one decimal fraction by another is performed as follows: Suppose it required to divide 6.42

by 1.213. The first of these is $\frac{642}{100}$, and the second $\frac{1233}{1000}$. The quotient of these by the ordinary rule is $\frac{642000}{121300}$, or $\frac{6420}{1213}$. This fraction must now be reduced to a decimal on the principles of the last article, by the rule usually given, either exactly, or by approximation, according to the nature of the factors in the denominator.

When the decimal fraction corresponding to a common fraction cannot be exactly found, it always happens that the series of decimals which approximates to it, contains the same number repeated again and again. Thus, in the example which we chose, $\frac{7}{11}$ is more and more nearly represented by the fractions .6, .63, .636, .6363, etc., and if we carried the process on without end, we should find a decimal fraction consisting entirely of repetitions of the figures 63 after the decimal point. Thus, in finding $\frac{1}{7}$, the figures which are repeated in the numerator are 142857. This is what is commonly called a circulating decimal, and rules are given in books of arithmetic for reducing them to common fractions. We would recommend to the beginner to omit all notice of these fractions, as they are of no practical use, and cannot be thoroughly understood without some knowledge of algebra. It is sufficient for the student to know that he can always either reduce a common fraction to a decimal, or find a decimal near enough to it for his purpose, though the calculation in which he is engaged requires a degree of accuracy which the finest microscope will not appreciate. But in using approximate decimals there is one remark of importance, the necessity for which occurs continually.

Suppose that the fraction 2.143876 has been obtained, and that it is more than sufficiently accurate for the calculation in which it is to be employed. Suppose that for the object proposed it is enough that each quantity employed should be a decimal fraction of three places only, the quantity 2.143876 is made up of 2.143, as far as three places of decimals are concerned, which at first sight might appear to be what we ought to use, instead of 2.143876. But this is not the number which will in this case give the utmost accuracy which three places of decimals will admit of; the common usages of life will guide us in this case. Suppose a regiment consists of 876 men, we should express this in what we call round numbers, which in this case would be done by saying how many hundred men there are, leaving out of consideration the number 76, which is not so great as 100; but in doing this we shall be nearer the truth if we say that the regiment consists of 900 men instead of 800, because 900 is nearer to 876 than 800. In the same manner, it will be nearer the truth to write 2.144 instead of 2.143, if we wish to express 2.143876 as nearly as possible by three places of decimals, since it will be found by subtraction that the first of these is nearer to the third than the second. Had the fraction been 2.14326, it would have been best expressed in three places by 2.143; had it been 2.1435, it would have been equally well expressed either by 2.143 or 2.144, both being equally near the truth; but 2.14351 is a little more nearly expressed by 2.144 than by 2.143.

We have now gone through the leading principles of arithmetical calculation, considered as a part of general Mathematics. With respect to the commercial rules, usually considered as the grand object of an arithmetical education, it is not within the scope of this treatise to enter upon their consideration. The mathematical student, if he is sufficiently well versed in their routine for the purposes of common life, may postpone their consideration until he shall have become familiar with algebraical operations, when he will find no difficulty in understanding the principles or practice of any of them. He should, before commencing the study of algebra, carefully review what he has learnt in arithmetic, particularly the reasonings which he has met with, and the use of the signs which have been introduced. Algebra is at first only arithmetic under another name, and with more general symbols, nor will any reasoning be presented to the student which he has not already met with in establishing the rules of arithmetic. His progress in the former science depends most materially, if not altogether, upon the manner in which he has attended to the latter; on which account the detail into which we have entered on some things which to an intelligent person are almost self-evident, must not be deemed superfluous.

When the student is well acquainted with the principles and practice of arithmetic, and not before, he should commence the study of algebra. It is usual to begin algebra and geometry together, and if the student has sufficient time, it is the best plan which he can adopt. Indeed, we see no reason why the elements of geometry should not precede those of algebra, and be studied together with arithmetic. In this casc the student should read some treatise which relates to geometry, first. It is hardly necessary to say that though we have adopted one particular order, yet the student may reverse or alter that order so as to suit the arrangement of his own studies.

We now proceed to the first principles of algebra, and the elucidation of the difficulties which are found from experience to be most perplexing to the beginner. We suppose him to be well acquainted with what has been previously laid down in this treatise, particularly with the meaning of the signs $+$, $-$, $\times$, and the sign of division.

CHAPTER VI.

ALGEBRAICAL NOTATION AND PRINCIPLES.

WHENEVER any idea is constantly recurring, the best thing which can be done for the perfection of language, and consequent advancement of knowledge, is to shorten as much as possible the sign which is used to stand for that idea. All that we have accomplished hitherto has been owing to the short and expressive language which we have used to represent numbers, and the operations which are performed upon them. The first step was to write simple signs for the first numbers, instead of words at full length, such as 8 and 7, instead of eight and seven. The next was to give these signs an additional meaning, according to the manner in which they were connected with one another; thus 187 was made to represent one hundred added to eight tens added to seven. The next was to give by new signs directions when to perform the operations of addition, subtraction, multiplication, and division; thus $5+8$ was made to represent 8 added to 5, and so on. With these signs reasonings were made, and truths discovered which are common to all numbers; not at once for every number, but by taking some example, by reasoning upon it, and by producing a result; this result led to a rule which was declared to be a rule which held equally good for all numbers, because the reasoning which produced it might have been applied to any other example as well as to the one which was chosen. In this way we produced some results, and might have produced many more; the following is an instance: half the sum of two numbers added to half their difference, gives the greater of the two numbers. For example, take 16 and 10, half their sum is 13, half their difference is 3; if we add 13 and 3 we get 16, the greater of the two numbers. We might satisfy ourselves of the truth of this same proposition for any other numbers, such as 27 and 8, 15 and 19, and so on. If we then make use of signs, we find the following truths:

$$\frac{16+10}{2}+\frac{16-10}{2}=16,$$
$$\frac{27+8}{2}+\frac{27-8}{2}=27,$$
$$\frac{15+9}{2}+\frac{15-9}{2}=15,$$

and so on. If, then, we choose any two numbers, and call them the first and second numbers, and call that the first number which is the greater of the two, we have the following:

$$\frac{\text{First No.} + \text{Second No.}}{2}+\frac{\text{First No.} - \text{Second No.}}{2}=\text{First No.}$$

In this way we might express anything which is true of all numbers, by writing First No., Second No., etc., for the different numbers which enter into our proposition, and we might afterwards suppose the First No., the Second No., etc., to be any which we please. In this way we might write down the following assertion, which we should find to be always true:

$$(\text{First No.} + \text{Second No.}) \times (\text{First No.} - \text{Second No.})$$
$$= \text{First No.} \times \text{First No.} - \text{Second No.} \times \text{Second No.}$$

When any sentence expresses that two numbers or collections of numbers are equal to one another, it is called an *equation** thus $7+5=12$ is an equation, and the sentences written just above are equations.

Now the next question is, could we not avoid the trouble of writing First No., Second No., etc., so frequently? This is done by putting letters of the alphabet to stand for these numbers. Suppose, e. g., we let x stand for the first number, and y for the second, the two assertions already made will then be written:

$$\frac{x+y}{2} + \frac{x-y}{2} = x,$$
$$(x+y) \times (x-y) = x \times x - y \times y.$$

By the use of letters we are thus enabled to write sentences which say something of all numbers, with a very small part only of the time and trouble necessary for writing the same thing at full length. We now proceed to enumerate the various symbols which are used.

1. The letters of the alphabet are used to stand for numbers, and whenever a letter is used it means either that any

*As now usually defined an *equation* always contains an unknown quantity. See also p. 90.—*Ed.*

number may be used instead of that letter, or that the number which the letter stands for is not known, and that the letter supplies its place in all the reasonings until it is known.

2. The sign + is used for addition, as in arithmetic. Thus $x+z$ is the sum of the numbers represented by x and z. The following equations are sufficiently evident:

$$x+y+z=x+z+y=y+z+x.$$

If $a=b$, then $a+c=b+c$, $a+c+d=b+c+d$, etc.

3. The sign $-$ is used for subtraction, as in arithmetic. The following equations will show its use:

$$\begin{aligned} x+a-b-c+e &= x+a+e-b-c \\ &= a-c+e-b+x. \end{aligned}$$

If $a=b$, then $a-c=b-c$, $a-c+d=b-c+d$, etc.

4. The sign $\times$ is used for multiplication as in arithmetic, but when two numbers represented by letters are multiplied together it is useless, since $a\times b$ can be represented by putting a and b together thus, ab. Also $a\times b\times c$ is represented by abc; $a\times a\times a$, for the present we represent thus, aaa. When two numbers are multiplied together, it is necessary to keep the sign $\times$; otherwise 7×5 or 35 would be mistaken for 75. It is, however, usual to place a point between two numbers which are to be multiplied together; thus $7\times 5\times 3$ is written $7\cdot 5\cdot 3$. But this point may sometimes be mistaken for the decimal point: this will, however, be avoided by always writing the

decimal point at the head of the figure, viz., by writing $\frac{23461}{100}$ thus, 234·61.

5. Division is written as in arithmetic: thus, $\frac{a}{b}$ signifies that the number represented by a is to be divided by the number represented by b.

6. All collections of numbers are called expressions; thus, $a + b$, $a + b - c$, $aa + bb - d$, are algebraical expressions.

7. When two expressions are to be multiplied together, it is indicated by placing them side by side, and inclosing each of them in brackets. Thus, if $a + b + c$ is to be multiplied by $d + e + f$, the product is written in any of the following ways:

$$(a + b + c)(d + e + f),$$
$$[a + b + c][d + e + f],$$
$$\{a + b + c\}\{d + e + f\},$$
$$\overline{a + b + c} \cdot \overline{d + e + f}.$$

8. That a is greater than b is written thus, $a > b$.

9. That a is less than b is written thus, $a < b$.

10. When there is a product in which all the factors are the same, such as $xxxxx$, which means that five equal numbers, each of which is represented by x, are multiplied together, the letter is only written once, and above it is written the number of times which it occurs, thus $xxxxx$ is written x^5. The following table should be carefully studied by

the student:

$x \times x$ or xx is written x^2,

and is called the square, or second power of x,

$x \times x \times x$ or xxx is written x^3,

and is called the cube, or third power of x,

$x \times x \times x \times x$ or $xxxx$ is written x^4,

and is called the fourth power of x,

$x \times x \times x \times x \times x$ or $xxxxx$ is written x^5,

and is called the fifth power of x,

etc., etc., etc.

There is no point which is so likely to create confusion in the ideas of a beginner as the likeness between such expressions as $4x$ and x^4. On this account it would be better for him to omit using the latter expression, and to put $xxxx$ in its place until he has acquired some little facility in the operations of algebra. If he does not pursue this course, he must recollect that the 4, in these two expressions, has different names and meanings. In $4x$ it is called a *coefficient*, in x^4 an *exponent* or *index*.

The difference of meaning will be apparent from the following tables:

$$\begin{array}{lll} 2x = x + x & x^2 = x \times x = xx, & \\ 3x = x + x + x & x^3 = x \times x \times x \quad \text{or} \quad xxx, & \\ 4x = x + x + x + x & x^4 = x \times x \times x \times x \quad \text{or} \quad xxxx, & \\ \qquad \text{etc.,} & \qquad \text{etc.} & \end{array}$$

$$\text{If } x = 3, \quad \begin{array}{ll} 2x = 6, & x^2 = 9, \\ 3x = 9, & x^3 = 27, \\ 4x = 12, & x^4 = 81. \end{array}$$

The beginner should frequently *write* for himself such expressions as the following:

$$4a^3b^2 = aaabb + aaabb + aaabb + aaabb,$$

$$5a^4x = aaaax + aaaax + aaaax + aaaax + aaaax,$$

$$9a^2b^3 + 4ab^4 = 9aabbb + 4abbbb,$$

$$\frac{a^2+b^2}{a^2-b^2} = \frac{aa+bb}{aa-bb} = \frac{aa}{aa-bb} + \frac{bb}{aa-bb}$$

$$= \frac{aa-cc}{aa-bb} + \frac{bb+cc}{aa-bb},$$

$$\frac{a^3-b^3}{a^2-b^2} = \frac{aaa-bbb}{aa-bb} = \frac{aa+ab+bb}{a+b}.$$

With many such expressions every book on algebra will furnish him, and he should then satisfy himself of their truth by putting some numbers at pleasure instead of the letters, and making the results agree with one another. Thus, to try the expression

$$\frac{a^3 - b^3}{a - b} = a^2 + ab + b^2,$$

or, which is the same,

$$\frac{aaa - bbb}{a - b} = a^2 + ab + b^2.$$

Let a stand for 6 and b stand for 4, then, if this expression be true,

$$\frac{6 \cdot 6 \cdot 6 - 4 \cdot 4 \cdot 4}{6 - 4} = 6 \cdot 6 + 6 \cdot 4 + 4 \cdot 4,$$

which is correct, since each of these expressions is found, by calculation, to be 76.

The student should then exercise himself in the solution of such questions as the following: What is

$$a^2 + b^2 - \frac{ab}{a+b} + \frac{a}{a+b} - a,$$

I. when a stands for 6, and b for 5, II. when a stands for 13, and b for 2, and so on. He should stop here until he has, by these means, made the signs familiar to his eye and their meaning to his mind; nor should he proceed to any further algebraical operations until he can readily find the value of any algebraical expression when he knows the numbers which the letters stand for. He cannot, at this period of his course, write too many algebraical expressions, and he must particularly avoid slurring over the sense of what he has before him, and must write and rewrite each expression until the meaning of the several parts forces itself upon his memory at first sight, without even the necessity of putting it in words. It is the neglecting to do this which renders the operations of algebra so tedious to the beginner. He usually proceeds to the addition, subtraction, etc., of symbols, of the meaning of which he has but an imperfect idea, and which have been newly introduced to him in such numbers that perpetual confusion is the consequence. We cannot, therefore, use too many arguments to induce him not to mind the drudgery of reducing algebraical expressions into figures. This is the connecting link between the new science and arithmetic, and, unless that link be well fastened, the knowledge which he has previously acquired in arithmetic will help him but little in acquiring algebra.

The order of the terms of any algebraical expression may be changed without changing the value of that expression. This needs no proof, and the following are examples of the change:

$$\begin{aligned} &a+b+ab+c+ac-d-e-de-f \\ &\qquad = a-d+b-e+ab-de+c-f+ac \\ &\qquad = a+b-d-e-de-f+ac+c+ab \\ &\qquad = ab+ac-de+a+b+c-e-f-d. \end{aligned}$$

When the first term changes its place, as in the fourth of

these expressions, the sign $+$ is put before it, and should, properly speaking, be written wherever there is no sign, to indicate that the term in question increases the result of the rest, but it is usually omitted. The negative sign is often written before the first term, as in the expression $-a+b$: but it must be recollected that this is written on the supposition that a is subtracted from what comes after it.

When an expression is written in brackets, with some sign before it, such as $a-(b-c)$, it is understood that the expression in brackets is to be considered as one quantity, and that its result or total is to be connected with the rest by the sign which precedes the brackets. In this example it is the *difference* of b and c which is to be subtracted from a. If $a=12$, $b=6$, and $c=4$, this is 10. In the expression $a-b$ made by subtracting b from a, too much has been subtracted by the quantity c, since it is not b, but $b-c$, which must be subtracted from a. In order, therefore, to make $a-(b-c)$, c must be added to $a-b$, which gives $a-b+c$. Therefore, $a-(b-c)=a-b+c$. Similarly

$$a+b-(c+d-e-f)=a+b-c-d+e+f,$$

$$(ax^2-bx+c)-(dx^2-ex+f)$$
$$=ax^2-bx+c-dx^2+ex-f.$$

When the positive sign is written before an expression in brackets, the brackets may be omitted altogether, unless they serve to show that the expression in question is multiplied by some other. Thus, instead of $(a+b+c)+(d+e+f)$, we may write $a+b+c+d+e+f$, which is, in fact, only saying that two wholes may be added together by adding together all the parts of which they are composed. But the expression $a+(b+c)(d+e)$ must not be written thus: $a+b+c(d+e)$, since the first expresses that $(b+c)$ must be multiplied by $(d+e)$ and the product added to a, and the second that c must be multiplied by $(d+e)$ and the product added to $a+b$. If a, b, c, d, and e, stand for 1, 2, 3, 4, and 5, the first is 46 and the second 30.

When two or more quantities have exactly the same letters repeated the same number of times, such as $4a^2b^3$, and $6a^2b^3$, they may be reduced into one by merely adding the coefficients and retaining the same letters. Thus, $2a+3a$ is $5a$, $6bc-4bc$ is $2bc$, $3(x+y)+2(x+y)$ is $5(x+y)$. These things are evident, but beginners are very liable to carry this farther than they ought, and to attempt to reduce expressions which do not admit of reduction. For example, they will say that $3b+^2$ is $4b$ or $4b^2$, neither of which is true, except when b stands for 1. The expression $3b+b^2$, or $3b+bb$, cannot be made more simple until we know what b stands for. The following table will, perhaps, be of service:

$$\begin{array}{ll} 6a^2b^3+3a^3b^2 & \text{is not } 9a^5b^5, \\ 6a^3-4a^2 & \text{is not } 2a, \\ 2ba+3b & \text{is not } 5ab. \end{array}$$

Such are the mistakes which beginners almost universally make, mostly for want of a moment's consideration. They attempt to reduce quantities which cannot be reduced, which they do by adding the exponents of letters as well as their coefficients, or by collecting several terms into one, and leaving out the signs of addition and subtraction. The beginner cannot too often repeat to himself that two terms can never be made into one, unless both have the same letters, each letter being repeated the same number of times in both, that is, having the same index in both. When this is the case, the expressions may be reduced by adding or subtracting the coefficients according to the sign, and affixing the common letters with their indices. When there is no coefficient, as in the expression a^2b, the quantity represented by a^2b being only taken once, 1 is called the coefficient. Thus,

$$3ab + 4ab + 6ab - ab - 7ab = 5ab,$$
$$6xy^2 + 3xy^2 - 5xy^2 + xy^2 = 5xy^2.$$

The student must also recollect that he is not at liberty to change an index from one letter to another, as by so doing he changes the quantity represented. Thus a^4b and ab^4 are quantities totally distinct, the first representing *aaaab* and the second *abbbb*. The difference in all the cases which we have mentioned will be made more clear, by placing numbers at pleasure instead of letters in the expressions, and calculating their values; but, in conclusion, the following remark must be attended to. If it were asserted that the expression $\dfrac{a^2+b^2}{a+b}$ is the same as $a+b-\dfrac{2ab}{2a-b}$, and we wish to proceed to see whether this is always the case or no, if we commence accidentally by supposing b to stand for 2 and a for 4, we shall find that the first is the same as the second, each being $3\frac{1}{2}$. But we must not conclude from this that they are always the same, at least until we have tried whether they are so, when other numbers are substituted for a and b. If we place 6 and 8 instead of a and b, we shall find that the two expressions are not equal, and therefore we must conclude that they are not always the same. Thus in the expressions $3x - 4$ and $2x + 8$, if x stand for 12, these are the same, but if it stands for any other number they are not the same.

CHAPTER VII.

ELEMENTARY RULES OF ALGEBRA.

THE student should be very well acquainted with the principles and notation hitherto laid down before he proceeds to the algebraical rules for addition and subtraction. He should then take some simple examples of each, and proceed to find the sum and difference by reasoning as follows. Suppose it is required to add $c - d$ to $a - b$. The direction to do this may either be written in the common way thus:

$$\begin{array}{rc} & a - b \\ & \underline{c - d} \\ \text{Add} & \end{array}$$

or more properly thus: Find $(a - b) + (c - d)$.

If we add c to a, or find $a + c$, we have too much; first, because it is not a which is to be increased by $c - d$ but $a - b$; this quantity must therefore be decreased by b on this account, or must become $a + c - b$; but this is still too great, because it is not c which was to be added but $c - d$; it must therefore be decreased by d on this account, or must become $a + c - b - d$ or $a - b + c - d$. From a few reasonings of this sort the rule may be deduced; and not till then should an example be chosen so complicated as to make the student lose sight for one moment of his demonstration. The process of subtraction we have already performed, and from a few examples of this method the rule may be deduced.

The multiplication of a by $c - d$ is performed thus: a is to be taken $c - d$ times. Take it first c times or find ac. This is too great, because a has been taken too many times by d. From ac we must therefore subtract d times a, or ad, and the result is that

$$a(c - d) = ac - ad.$$

This may be verified from arithmetic, in which the same process is shown to be correct; and this whether the numbers a, c, and d are whole or fractional. For example, it will be found that $6(14 - 9)$ or 6×5 is the same as $6 \times 14 - 6 \times 9$, or as $84 - 54$. Also that $\frac{2}{3}(\frac{1}{7} - \frac{2}{15})$, or $\frac{2}{3} \times \frac{1}{105}$ is the same as $\frac{2}{3} \times \frac{1}{7} - \frac{2}{3} \times \frac{2}{15}$, or as $\frac{2}{21} - \frac{4}{45}$. Upon similar reasoning the following equations may be proved:

$$a(b + c - d) = ab + ac - ad,$$
$$(p + pq - ar)xz = pxz + pqxz - arxz,$$
$$(a^2 + 2b^2)b^2, \quad \text{or} \quad (aa + 2bb)bb = aabb + 2bbbb = a^2b^2 + 2b^4.$$

Also when a multiplication has been performed, the process may be reversed and the factors of it may be given. Thus, on observing the expression

$$ab - ac + a^2,$$
$$\text{or} \quad ab - ac + aa,$$

we see that in its formation every term has been multiplied

by a; that is, it has been made by multiplying

$$b - c + a \text{ by } a,$$
$$\text{or } a \text{ by } b - c + a.$$

There will now be no difficulty in perceiving that

$$\begin{aligned} ac + ad + bc + bd &= a(c + d) + b(c + d) \\ &= (a + b)(c + d), \\ a^2 - ab^2 + 2abc - dc + 3a &= a(a - b^2 + 3) + c(2ba - d). \end{aligned}$$

It is proved in arithmetic that if numbers, whether whole or fractional, are multiplied together, the product remains the same when the order in which they are multiplied is changed. Thus $6 \times 4 \times 3 = 3 \times 6 \times 4 = 4 \times 6 \times 3$, etc., and $\frac{2}{3} \times \frac{4}{5} = \frac{4}{5} \times \frac{2}{3}$, etc. Also, that a part of the multiplication may be made, and the partial product substituted instead of the factors which produced it, thus, $3 \times 4 \times 5 \times 6$ is $12 \times 5 \times 6$, or $15 \times 4 \times 6$, or 90×4. From these rules two complicated single terms may be multiplied together, and the product represented in the most simple manner which the case admits of. Thus if it be required to multiply

$$6a^3b^4c, \text{ which is } 6\,aaa\,bbbb\,c$$
$$\text{by } 12a^2b^3c^3d, \text{ which is } 12\,aa\,bbb\,ccc\,d,$$

the product is written thus:

$$6\,aaa\,bbbb\,c\,12\,aa\,bbb\,ccc\,d,$$

which multiplication may be performed in the following order

$$6 \times 12\, aaaaa\, bbbbbbb\, cccc\, d,$$

which is represented by $72a^5b^7c^4d$. A few examples of this sort will establish the rule for the multiplication of such quantities which is usually given in the treatises on Algebra.

It is to be recollected that in every algebraical formula which is true of all numbers, any algebraical expression may be substituted for one of the letters, provided care is taken to make the substitution wherever that letter occurs. Thus from the formula:

$$a^2 - b^2 = (a+b)(a-b)$$

we may deduce the following by making substitutions for a. If this formula be always true, it is true when a is equal to $p+q$, that is, it is true if $p+q$ be put instead of a wherever that letter occurs in the formula. Therefore,

$$(p+q)^2 - b^2 = (p+q+b)(p+q-b).$$

Similarly,

$$(b+m)^2 - b^2 = (2b+m)m,$$

$$(x+y)^2 - (x-y)^2 = (x+y+x-y)(x+y-\overline{x-y}) = 4xy,$$

and so on.

We have already established the formula,

$$(p - q)a = ap - aq.$$

Instead of a let us put $r - s$, and this formula becomes

$$(p - q)(r - s) = (r - s)p - (r - s)q.$$

But

$$(r - s)p = pr - ps, \quad \text{and} \quad (r - s)q = qr - qs.$$

Therefore

$$\begin{aligned}(p - q)(r - s) &= pr - ps - (qr - qs) \\ &= pr - ps - qr + qs.\end{aligned}$$

By reasoning in the same way we may prove that

$$(p - q)(r + s) = pr + ps - qr - qs.$$

A few examples of this sort will establish what is called the rule of signs in multiplication; viz., that a term of the multiplicand multiplied by a term of the multiplier has the sign + before it if the terms have the same sign, and − if they have different signs. But here the student must avoid using an incorrect mode of expression, which is very common, viz., the saying that + multiplied by + gives +; − multiplied by + gives −; and so on. He must recollect that the signs + and − are not quantities, but *directions* to add and subtract, and that, as has been well said by one of the most luminous writers on algebra in our language, we might as well say, that take away multiplied by take away gives add, as that − multiplied by − gives +.*

The only way in which the student should accustom himself to state this rule is the following: "In multiplying two algebraical expressions, multiply each term of the one by each term of the other, and wherever two terms are preceded by the same sign put + before the product of the two; when the signs are different put the sign − before their product."

If the student should meet with an equation in which positive and negative signs stand by themselves, such as

$$+ab \times -c = -abc,$$

let him, for the present, reject the example in which it occurs, and defer the consideration of such equations until he has read the explanation of them to which we shall soon come. Above all, he must reject the definition still sometimes given of the quantity $-a$, that it is less than nothing. It is astonishing that the human intellect should ever have tolerated such an absurdity as the idea of a quantity less than nothing;† above all, that the notion should have outlived the

*Frend, *Principles of Algebra.* The author of this treatise is far from agreeing with the work which he has quoted in the rejection of the isolated negative sign which prevails throughout it, but fully concurs in what is there said of the methods then in use for explaining the difficulties of the negative sign.

†For a full critical and historical discussion of this point, see

belief in judicial astrology and the existence of witches, either of which is ten thousand times more possible.

These remarks do not apply to such an expression as $-b+a$, which we sometimes write instead of $a-b$, as long as it is recollected that the one is merely used to stand for the other, and for the present a must be considered as greater than b.

In writing algebraical expressions, we have seen that various arrangements may be adopted. Thus $ax^2 - bx + c$ may be written as $c + ax^2 - bx$, or $-bx + c + ax^2$. Of these three the first is generally chosen, because the highest power of x is written first; the highest but one comes next; and last of all the term which contains no power of x. When written in this way the expression is said to be arranged in descending powers of x; had it been written thus, $c - bx + ax^2$, it would have been arranged in ascending powers of x; in either case it is said to be arranged in powers of x, which is called the principal letter. It is usual to arrange all expressions which occur in the same question in powers of the same letter, and practice must dictate the most convenient arrangement. Time and trouble is saved by this operation, as will be evident from multiplying two unarranged expressions together, and afterwards doing the same with the same expressions properly arranged.

In multiplying two arranged expressions together, while

Duhamel. *Des méthodes dans les sciences de raisonnement*, 2me partie, chap. XIX. (third edition, Paris, Gauthier-Villars, 1896).—*Editor.*

collecting such terms into one as will admit of it, it will always be evident that the first and last of all the products contain powers of the principal letter which are found in no other part, and stand in the product unaltered by combination with any other terms, while in the intermediate products there are often two or more which contain the same power of the principal letter, and can be reduced into one. This will be evident in the following examples:

$$\begin{array}{ll}
\text{Multiply} & x^6 - 3x^5 + x^4 \\
\quad\text{By} & x^4 - 2x^2 + x \\
\hline
\text{The product is} & x^{10} - 3x^9 + x^8 \\
 & \qquad\quad - 2x^8 + 6x^7 - 2x^6 \\
 & \qquad\qquad\qquad + x^7 - 3x^6 + x^5 \\
\hline
\text{Or} & x^{10} - 3x^9 - x^8 + 7x^7 - 5x^6 + x^5.
\end{array}$$

$$\begin{array}{ll}
\text{Multiply} & ax^3 + bx^2 + cx \\
\quad\text{By} & dx^2 + ex + f \\
\hline
\text{The product is} & adx^5 + bdx^4 + cdx^3 \\
 & \qquad + aex^4 + bex^3 + cex^2 \\
 & \qquad\qquad + afx^3 + bfx^2 + cfx \\
\hline
\text{Or} & adx^5 + (bd + ae)x^4 \\
 & \qquad + (cd + be + af)x^3 \\
 & \qquad\qquad + (ce + bf)x^2 \\
 & \qquad\qquad\qquad + cfx.
\end{array}$$

It is plain from the rule of multiplication, that the highest power of x in a product must be formed by multiplying the highest power in one factor by the highest power in the other, or when the two factors have been arranged in descending powers, the *first* power in one by the first power in the other. Also, that the lowest power of x, or should it so happen, the term in which there is no power of x, is made by multiplying the last terms in each factor. These being the highest and lowest, there can be no other such power, consequently neither of these terms can coalesce with any other, as is the case in the intermediate products. This remark will be of most convenient application in division, to which we now come.

Division is in all respects the reverse of multiplication. In dividing a by b we find the answer to this question: If a be divided into b equal parts, what is the magnitude of each of those parts? The quotient is, from the definition of a fraction, the same as the fraction $\frac{a}{b}$, and all that remains is to see whether that fraction can be represented by a simple algebraical expression without fractions or not; just as in arithmetic the division of 200 by 26 is the reduction of the fraction $\frac{200}{26}$ to a whole number, if possible. But we must here observe that a distinction must be drawn between algebraical and arithmetical fractions. For example, $\frac{a+b}{a-b}$ is an algebraical fraction, that is, there is no expression without fractions which is always equal to $\frac{a+b}{a-b}$. But it does not follow from this that the number which $\frac{a+b}{a-b}$ represents is always an arithmetical fraction; the contrary may be shown. Let a stand for 12, and b for 6, then $\frac{a+b}{a-b}$ is 3. Again, a^2+ab is a quantity which does not contain algebraical fractions, but it by no means follows that it may not represent an arithmetical fraction. To show that it may, let $a=\frac{1}{2}$ and $b=2$, then $a^2+ab=1\frac{1}{4}$ or $\frac{5}{4}$. Other examples will clear up this point if any doubt yet exist in the mind of the student. Nevertheless, the following propositions of arithmetic and algebra, which only differ in this, that "*whole number*" in the arithmetical proposition is replaced by "*simple expression*"* in the algebraical one, connect the two subjects and render those demonstrations which are in arithmetic confined to whole numbers, equally true in algebra as far as regards simple expressions:

The sum, difference, or product of two whole numbers, is a whole number.	The sum, difference, or product of two simple expressions is a simple expression.
One number is said to be a measure of another when the quotient of the two is a whole number.	One expression is said to be a measure of another when the quotient of the two is a simple expression.

*By a simple expression is meant one which does not contain the principal letter in the denominator of any fraction.

The greatest common measure of two whole numbers is the greatest whole number which measures both, and is the product of all the prime numbers which will measure both.

The greatest common measure of two expressions is the common measure which has the highest exponents and coefficients, and is the product of all prime simple expressions which measure both.

When one number measures two others, it measures their sum, difference, and product.

When one expression measures two others, it measures their sum, difference, and product.

In the division of one number by another, the remainder is measured by any number which measures the dividend and divisor.

In the division of one expression by another, the remainder is measured by any expression which measures the dividend and divisor.

A fraction is not altered by multiplying or dividing both its numerator and denominator by the same quantity.

A fractional expression is not altered by multiplying or dividing both its numerator and denominator by the same expression.

In the term *simple expression* are included those quantities which contain arithmetical fractions, provided there is no algebraical quantity, or quantity represented by letters in the denominator; thus $\frac{1}{4}ab + \frac{1}{2}$ is called a simple expression. We now proceed to the division of one simple expression by another, and we will take first the case where neither quantity contains more than one term. For example, what is

$42a^4b^3c$ divided by $6a^2bc$? that is, what quantity must be multiplied by $6a^2bc$, in order to produce $42a^4b^3c$. This last expression written at length, is $42\,aaaa\,bbb\,c$, and 42 is 6×7. We can then separate this expression into the product of two others, one of which shall be $6a^2bc$, or $6aabc$; it will then be $6aabc \times 7aabb$, and it is $7aabb$ which must be multiplied by $6aabc$ in order to produce $42a^4b^3c$. A few examples worked in this way, will lead the student to the rule usually given in all cases but one, to which we now come. We have represented cc, ccc, $cccc$, etc., by c^2, c^3, c^4, etc., and have called them the second, third, fourth, etc., powers of c. The extension of this rule would lead us to represent c by c^1, and call it the first power of c. Again, we have represented $c+c$, $c+c+c$, $c+c+c+c$, etc. by $2c$, $3c$, $4c$, and have called 2, 3, 4, etc., the coefficients of c. The extension of this rule would lead us to write c thus, $1c$, or, rather, if we attend to the last remark, $1c^1$. This instance leads us to observe the gradual progress of our language. We begin with the quantity c by itself; we proceed in our course, shortening by new signs the more complicated combinations of c, and the original quantity c forces itself anew upon our attention as a part of the series,

$$c,\ 2c,\ 3c,\ 4c, \text{ etc., and } c,\ c^2,\ c^3,\ c^4, \text{ etc.,}$$

in each of which, except the first, there is a distinct figure, which is called a coefficient or exponent, according to its situation. We then deduce rules in which the terms coefficient or exponent occur, but which, of course, cannot apply to the first term in each series, because, as yet, it has neither coefficient nor exponent. Among such rules are the following:

I. To add two terms of the first series, add the coefficients, and affix to the sum the letter c. Thus $4c + 3c = 7c$.

II. To multiply two terms of the second series, add the exponents, and make this sum the exponent of c. Thus $c^4 \times c^3 = c^7$.

III. To divide a term of the second series by one which comes before it, subtract the exponent of the divisor from the exponent of the dividend, and make this difference the exponent of c. Thus,

$$\frac{c^7}{c^4} = c^3.$$

These rules are intelligible for all terms of the series except the first, to which, nevertheless, they will apply if we agree that $1c^1$ shall represent c, as will be evident by applying either of them to find $4c+c$, $c^4 \times c$, or $\frac{c^4}{c}$. We therefore *agree* that $1c^1$ shall stand for c, and although c is not written thus, it must be remembered that c is to be considered as having the coefficient 1 and the exponent 1, which is an amendment and enlargement of our algebraical language, derived from experience. It may be said that this is all superfluous, because, if c^2 stand for cc, and c^3 for ccc, what can c^1 stand for but c? But it must be recollected that, since the symbol c^1 has not yet received a meaning, we are at liberty to make it stand for anything which we please, for example, for $\frac{1+c}{c}$, or $c-c^2$, or any other. If we did this, there would, indeed, be a great violation of analogy, that is, what c^1 stands for would not be as like that which c^2 has been made to stand for, as the meaning of c^3 is to that of c^4; but, nevertheless, we should not be led to any incorrect results as long as we remembered to make c^1 always stand for the same thing. These remarks are here introduced in order to show the manner in which analogy is followed in extending the language of algebra, and to prove that, after a certain period, we may rather be said to discover new symbols than to make them. The immense importance of this branch of the subject makes it necessary that it should be fully and early understood by all who intend to pursue their mathematical studies to any depth. To illustrate it still further, we subjoin another instance, which has not been noticed in its proper place.

The signs $+$ and $-$ were first used to connect one quantity with others, and to show what arithmetical operations were performed on other quantities by means of the first. But the first quantity on which we begin the operation is not preceded by any sign, not being considered as added to or subtracted from any previous one. Rules were afterwards deduced for the addition and subtraction of the total result of several expressions in which these signs occur, as follows:

To add two expressions, form a third, which has all the quantities in the first two, with the same signs.

To subtract one expression from another, change the sign of each term of the subtrahend, and proceed as in the last rule.

The only terms in which these rules do not apply are those which have no sign, viz., the first of each. But they will apply to those terms, and will produce correct results, if we place the sign + before each of them. We are thus led to see that an algebraical term which has no sign is equivalent in all operations to one which is preceded by the sign +. We, therefore, consider this sign as prefixed, though it is not always written, and thus we are furnished with a method of containing under one rule that which would otherwise require two.

From these considerations the following appears to be the best and most natural course of proceeding in the invention of additional symbols. When a rule has been discovered which is not quite general, and which only fails in its application to a few instances, annex such additional symbols to those already in use, or change and modify these so as to make the rule applicable in all cases, provided always this can be done without making the same symbol stand for two different things, and without any violation of analogy. If the rule itself, by its application to any case, should produce a new symbol hitherto unexplained, it is a sign that the rule has been applied to a case which was never intended to fall under it when it was made. For the solution of this case we must have recourse to first principles, but when, by these means, the result has been found, it will be best to agree that the new symbol furnished by the rule shall stand for the result furnished by the principle, by which means the generality of the rule will be attained and the analogy of language will not be injured. Of this the following is a remarkable instance:

To divide c^8 by c^5 the rule tells us to subtract 5 from 8, and make the result the exponent of c, which gives the quotient c^3. If we *apply the same rule* to divide c^6 by c^6, since 6 subtracted from 6 leaves 0, the result is c^0, a new symbol, to which we have attached no meaning. The fact is that the rule was formed from observation of different powers of c, and was never intended to apply to the division of a power of c by the same power. If we apply the common principles to the division of c^6 by c^6, the result is 1. We, therefore, agree that c^0 shall stand for 1, and the least inspection will show that this agreement does not affect the truth of any result derived from the rule. If, in the solution of any problem, the symbol c^0 should appear, we must consider it is a sign that we have, in the course of the investigation, divided a power of c by itself by the common rule, without remarking that the quotient is 1. We must, therefore, replace c^0 by 1, but it is entirely indifferent at what stage of the process this is done.

Several extensions might be noticed, which are made almost intuitively, to which these observations will apply. Such, for example, is the multiplication and division of any number by 1, which is not contemplated in the definition of these operations. Such is also the continual use of 0 as a quantity, the addition and subtraction of it from other quantities, and the multiplication of it by others, neither of which were contemplated when these operations were first thought of.

We now proceed to the principles on which more complicated divisions are performed. The question proposed in division, and the manner of answering it, may be explained in the following manner. Let A be an expression which is to be divided by H, and let Q be the quotient of the two. By the meaning of division, if there be no remainder $A = QH$, since the quotient is the expression which must multiply the divisor, in order to produce the dividend. Now let the quotient be made up of different terms, a, b, c, etc., let it be $a + b - c + d$. That is, let

$$A = QH, \tag{1}$$

$$Q = a + b - c + d. \tag{2}$$

By putting, instead of Q in (1), that which is equal to it in (2), we find

$$A = (a + b - c + d)H = aH + bH - cH + dH. \tag{3}$$

Now suppose that we can by any method find the term a of the quotient, that is, that we can by trial or otherwise find one term of the quotient. In (3), when the term a is found, since H is known, the term aH is found. Now if two quantities are equal, and from them we subtract the same quantity, the remainders will be equal. Subtract aH from the equal quantities A and $aH + bH - cH + dH$, and we shall find

$$A - aH = bH - cH + dH = (b - c + d)H. \tag{4}$$

If, then, we multiply the term of the quotient found by the divisor, and subtract the product from the dividend, and call the remainder B; then

$$B = (b - c + d)H. \tag{5}$$

That is, if B be made a dividend, and H still continue the divisor, the quotient is $b - c + d$, or all the first quotient, except the part of it which we have found. We then proceed in the same manner with this new dividend, that is, we find b and also bH, and subtract it from B, and let $B - bH$ be represented by C, which gives by the process which has just been explained

$$C = (-c + d)H = -cH + dH. \tag{6}$$

We now come to a negative term of the quotient. Let us suppose that we have found c, and that its sign in the quotient is $-$. If two quantities are equal, and we add the same quantity to both, the sums are equal. Let us therefore add cH to both the equal quantities in (6), and the equation will become

$$C + cH = dH; \tag{7}$$

or if we denote $C + cH$ by D, this is

$$D = dH.$$

There is only one term of the quotient remaining, and if that can be found the process is finished. But as we cannot know when we have come to the last term, we must continue the same process, that is, subtract dH from D, in doing which we shall find that dH is equal to D, or that the remainder is nothing. This indicates that the quotient is now exhausted and that the process is finished.

We will now apply this to an example in which the quotient is of the same form as that in the last process, namely, consisting of four terms, the third of which has the negative sign. This is the division of

$$x^4 - y^4 - 3x^2y^2 + x^3y + 2xy^3 \quad \text{by} \quad x - y.$$

Arrange the first quantity in descending powers of x which will make it stand thus:

$$x^4 + x^3y - 3x^2y^2 + 2xy^3 - y^4. \qquad (A)$$

One term of the quotient can be found immediately, for since it has been shown that the term containing the highest power of x in a product is made up of nothing but the product of the terms containing the highest powers of x which occur in the multiplier and multiplicand, and considering that the expression (A) is the product of $x - y$ and the quotient, we shall recover the highest power of x in the quotient by dividing x^4, the highest power of x in (A), by x, its highest power in $x - y$. This division gives x^3 as the first term of the quotient. The following is the common process, and with each line is put the corresponding step of the process above explained, of which this is an example:

$$
\begin{array}{llr}
 & (H) \qquad\qquad (A) & (a) \\
 & x - y)\ x^4 + x^3y - 3x^2y^2 + 2xy^3 - y^4 & (x^3 \\
(aH) & \text{Subtract } x^4 - x^3y & \\
\hline
 & & (b) \\
(B)\ \text{Second dividend} & 2x^3y - 3x^2y^2 + 2xy^3 - y^4 & (+2x^2y \\
(bH)\ \text{Subtract} & 2x^3y - 2x^2y^2 & \\
\hline
 & & (c) \\
(C)\ \text{Third dividend} & - x^2y^2 + 2xy^3 - y^4 & (-xy^2 \\
(cH)\ \text{Subtract} & - x^2y^2 + xy^3 & \\
\hline
 & & (d) \\
(D)\ \text{Fourth dividend} & xy^3 - y^4 & (+y^3 \\
(dH)\ \text{Subtract} & xy^3 - y^4 & \\
\hline
 & 0 &
\end{array}
$$

The whole quotient is therefore $x^3 + 2x^2y - xy^2 + y^3$.

The second and following terms of the quotient are determined in exactly the same manner as the first. In fact, this process is not the finding of a quotient directly from the divisor and dividend, but one term is first found, and by means of that term another dividend is obtained, which only differs from the first in having one term less in the quotient, viz., that which was first found. From this second dividend one term of its quotient is found, and so on until we obtain a dividend whose quotient has only one term, the finding of which finishes the process. It is usual also to neglect all the terms of the first dividend, except those which are immediately wanted, taking down the others one by one as they become necessary. This is a very good method in practice but should be avoided in explaining the principle, since the first subtraction is made from the whole dividend, though the operation may only affect the form of some part of it.

If the student will now read attentively what has been said on the greatest common measure of two numbers, and then examine the connexion of whole numbers in arithmetic and simple expressions in algebra with which we commenced the subject of division, he will see that the greatest algebraical common measure of two expressions may be found in exactly the same manner as the same operation is performed in arithmetic. He must also recollect that the greatest common measure of two expressions A and B is not altered by multiplying or dividing either of them, A, for example, by any quantity, provided that quantity has no measure in common with B. For example, the greatest common measure of $a^2 - x^2$ and $ba^3 - bx^3$ is the same with that of $2a^2 - 2x^2$ and $a^3 - x^3$, since though a new measure is now introduced into the first and taken away from the second, nothing is introduced or taken away which is common to both. The same observation applies to arithmetic also. For example, take the numbers 162 and 180. We may, without altering their greatest common measure, multiply the first by 7 and the second by 11, etc. The rule for finding the greatest common measure should be practised with great attention by all who intend to proceed beyond the usual stage in algebra. To others it is not of the same importance, as the necessity for it never occurs in the lower branches of the science.

In proceeding to the subject of fractions, it must be observed that, in the same manner as in arithmetic, when there is a remainder which cannot be further divided by the divisor, that is, where the dividend is so reduced that no simple term multiplied by the first term of the divisor will give the first term of the remainder, as in the case where the divisor is $a^2x + bx^2$ and the remainder $ax + b$; in this case a fraction must be added to the quotient, whose numerator is this remainder, and whose denominator is the divisor. Thus, in dividing $a^4 + b^4$ by $a + b$, the quotient is $a^3 - a^2b + ab^2 - b^3$, and the remainder $2b^4$, whence

$$\frac{a^4 + b^4}{a + b} = a^3 - a^2b + ab^2 - b^3 + \frac{2b^4}{a + b}.$$

The arithmetical rules for the addition, etc., of fractions hold equally good when the numerators and denominators are themselves fractions. Thus $\frac{\frac{3}{4}}{\frac{2}{7}}$ and $\frac{\frac{1}{5}}{\frac{3}{2}}$ are added, etc., exactly in the same way as $\frac{2}{5}$ and $\frac{3}{7}$, the sum of the second being

$$\frac{7 \times 2 + 5 \times 3}{5 \times 7}$$

and that of the first

$$\frac{\frac{3}{2} \times \frac{3}{4} + \frac{2}{7} \times \frac{1}{5}}{\frac{2}{7} \times \frac{3}{2}}.$$

The rules for the addition, etc., of algebraic fractions are exactly the same as in arithmetic; for both the numerator and denominator of every algebraic fraction stands either for a whole number or a fraction, and therefore the fraction itself is either of the same form as $\frac{6}{7}$ or $\frac{\frac{2}{3}}{\frac{4}{5}}$. Nevertheless the student should attend to some examples of each operation upon algebraic fractions, by way of practice in the previous operations. As the subject is not one which presents any peculiar difficulties, we shall now pass on to the subject of equations, concluding this article with a list of formulas which it is highly desirable that the student should commit to memory before proceeding to any other part of the subject.

$$(a+b)+(a-b)=2a, \tag{1}$$

$$(a+b)-(a-b)=2b, \tag{2}$$

$$a-(a-b)=b, \tag{3}$$

$$(a+b)^2=a^2+2ab+b^2, \tag{4}$$

$$(a-b)^2=a^2-2ab+b^2, \tag{5}$$

$$(2ax+b)^2=4a^2x^2+4abx+b^2, \tag{6}$$

$$(a+b)(a-b)=a^2-b^2, \tag{7}$$

$$\left.\begin{aligned}(x+a)(x+b)&=x^2+(a+b)x+ab,\\(x-a)(x-b)&=x^2-(a+b)x+ab,\end{aligned}\right\} \tag{8}$$

$$\frac{a}{b}=\frac{ma}{mb}, \tag{9}$$

$$a+\frac{c}{d}=\frac{ad+c}{d}, \qquad a-\frac{c}{d}=\frac{ad-c}{d}, \tag{10}$$

$$\frac{a}{b}+\frac{c}{d}=\frac{ad+bc}{bd}, \qquad \frac{a}{b}-\frac{c}{d}=\frac{ad-bc}{bd}, \tag{11}$$

$$\frac{a}{b}\times c=\frac{ac}{b}=\frac{a}{\frac{b}{c}}, \qquad \frac{a}{b}\times\frac{c}{d}=\frac{ac}{bd}, \tag{12}$$

$$\frac{\frac{a}{b}}{c}=\frac{a}{bc}=\frac{\frac{a}{c}}{b}, \tag{13}$$

$$\frac{\frac{a}{b}}{\frac{c}{d}}=\frac{ad}{bc}=\frac{\frac{a}{c}}{\frac{b}{d}}, \tag{14}$$

$$\frac{1}{\frac{a}{b}}=\frac{b}{a}. \tag{15}$$

CHAPTER VIII.

EQUATIONS OF THE FIRST DEGREE.

WE have already defined an equation, and have come to many equations of different sorts. But all of them had this character, that they did not depend upon the particular number which any letter stood for, but were equally true, whatever numbers might be put in place of the letters. For example, in the equation

$$\frac{a^2-1}{a+1}=a-1$$

the truth of the assertion made in this algebraical sentence is the same, whether a be considered as representing 1, 2, $2\frac{1}{2}$, etc., or any other number or fraction whatever. The second side of this equation is, in fact, the result of the operation pointed out on the first side. On the first side you are directed to divide a^2-1 by $a+1$; the second side shows you the result of that division. An equation of this description is called an *identical* equation, because, in fact, its two sides are but different ways of writing down the same number. This will be more clearly seen in the identical equations

$$a+a=2a, \quad 7a-3a+b=4a-3b+4b, \quad \text{and} \quad \frac{a}{b}\times b=a.$$

The whole of the formulæ at the end of the last article are examples of identical equations. There is not one of them which is not true for all values which can be given to the letters which enter into them, provided only that whatever a letter stands for in one part of an equation, it stands for the same in all the other parts.

If we take, now, such an equation as $a+1=8$, we have an equation which is no longer true for every value which can be given to its algebraic quantities. It is evident that the only number which a can represent consistently with this equation is 7, as any other supposition involves absurdity. This is a new species of equation, which can only exist in some particular case, which particular case can be found from the equation itself. The solution of every problem leads to such an equation, as will be shown hereafter, and, in the elements of algebra, this latter species of equation is of most importance. In order to distinguish them from identical equations, they are called *equations of condition*, because they cannot be true when the letters contained in them stand for any number whatever, and their very existence makes a condition which the letters contained must fulfil. The solution of an equation of condition is the process of finding what number the letter must stand for in order that the equation may be true. Every such solution is a process of reasoning, which, setting out with supposing the truth of the equation, proceeds by self-evident steps, making use of the common rules of arithmetic and algebra. We shall return to the subject of the solution of equations of condition, after showing, in a few instances, how we come to them in the solution of problems. In equations of condition, the quantity whose value is determined by the equation is usually represented by one of the last letters of the alphabet, and all others by some of the first. This distinction is necessary only for the beginner; in time he must learn to drop it, and consider any letter as standing for a quantity known or unknown, according to the conditions of the problem.

In reducing problems to algebraical equations no general rule can be given. The problem is some property of a number expressed in words by which that number is to be found, and this property must be written down as an equation in the most convenient way. As examples of this, the reduction of the following problems into equations is given:

I. What number is that to which, if 56 be added, the result will be 200 diminished by twice that number?

Let x stand for the number which is to be found.

Then $x + 56 = 200 - 2x$.

If, instead of 56, 200, and 2, any other given numbers, a, b, and c, are made use of in the same manner, the equation which determines x is

$$x + a = b - cx.$$

II. Two couriers set out from the same place, the second of whom goes three miles an hour, and the first two. The first has been gone four hours, when the second is sent after him. How long will it be before he overtakes him?

Let x be the number of hours which the second must travel to overtake the first. At the time when this event takes place, the first has been gone $x + 4$ hours, and will have travelled $(x+4)2$, or $2x+8$ miles. The second has been gone x hours, and will have travelled $3x$ miles. And, when the second overtakes the first, they have travelled exactly the same distance, and, therefore,

$$3x = 2x + 8.$$

If, instead of these numbers, the first goes a miles an hour, the second b, and c hours elapse before the second is sent after the first,

$$bx = ax + ac.$$

Four men, A, B, C, and D, built a ship which cost £2607, of which B paid twice as much as A, C paid as much as A and B, and D as much as C and B. What did each pay?

Suppose that A paid x pounds,
then B paid $2x$...,
C paid $x + 2x$ or $3x$...,
D paid $2x + 3x$ or $5x$....

All together paid $x + 2x + 3x + 5x$, or $11x$, therefore

$$11x = 2607.$$

There are two cocks, from the first of which a cistern is filled in 12 hours, and the second in 15. How long would they be in filling it if both were opened together?

Let x be the number of hours which would elapse before it was filled. Then, since the first cock fills the cistern in 12 hours, in one hour it fills $\frac{1}{12}$ of it, in two hours $\frac{2}{12}$, etc., and in x hours $\frac{x}{12}$. Similarly, in x hours, the second cock fills $\frac{x}{15}$ of the cistern. When the two have exactly filled the cistern, the sum of these fractions must represent a whole or 1, and, therefore,

$$\frac{x}{12} + \frac{x}{15} = 1.$$

If the times in which the two can fill the cistern are a and b hours, the equation becomes

$$\frac{x}{a} + \frac{y}{b} = 1.$$

A person bought 8 yards of cloth for £3 2*s.*, giving 9*s.* a yard for some of it and 7*s.* a yard for the rest; how much of each sort did he buy?

Let x be the number of yards at 7*s.* Then $7x$ is the number of shillings they cost. Also $8 - x$ is the number of yards at 9*s.*, and $(8-x)9$, or $72-9x$, is the number of shillings they cost. And the sum of these, or $7x+72-9x$, is the whole price, which is £3 2*s.*, or 62 shillings, and, therefore,

$$7x + 72 - 9x = 62.$$

These examples will be sufficient to show the method of reducing a problem to an equation. Assuming a letter to stand for the unknown quantity, by means of this letter the same quantity must be found in two different forms, and these must be connected by the sign of equality. However, the reduction into equations of such problems as are usually given in the treatises on algebra rarely occurs in the applications of mathematics. The process is a useful exercise of ingenuity, but no student need give a great deal of time to it. Above all, let no one suppose, because he finds himself unable to reduce to equations the conundrums with which such books are usually filled, that, therefore, he is not made for the study of mathematics, and should give it up. His future progress depends in no degree upon the facility with which he discovers the equations of problems; we mean as far as power of comprehending the subsequent sciences is concerned. He may never, perhaps, make any considerable step for himself, but, without doing this, he may derive all the benefits which the study of mathematics can afford, and even apply them extensively. There is nothing which discourages beginners more than the difficulty of reducing problems to equations, and yet, as respects its utility, if there be anything in the elements which may be dispensed with, it is this. We do not wish to depreciate its utility as an exercise for the mind, or to hinder all from attempting to conquer the difficulties which present themselves; but to remind every one that, if he can read and understand all that is set before him, the essential benefit derived from mathematical studies will be gained, even though he should never make one step for himself in the solution of any problem.

We return now to the solution of equations of condition.

Of these there are various classes. Equations of the first degree, commonly called simple equations, are those which contain only the first power of the unknown quantity. Of this class are all the equations to which we have hitherto come in the solution of problems. The principle by which they are solved is, that two equal quantities may be increased or diminished, multiplied, or divided by any quantity, and the results will be the same. In algebraical language, if $a = b$, $a + c = b + c$, $a - c = b - c$, $ac = bc$, and $\frac{a}{c} = \frac{b}{c}$. In every elementary book it is stated that any quantity may be removed from one side of the equation to the other, provided its sign be changed. This is nothing but an application of the principle just stated, as may be shown thus: Let $a+b-c = d$, add c to both quantities, then

$$a + b - c + c = d + c \quad \text{or} \quad a + b = d + c.$$

Again subtract b from both quantities, then $a + b - c - b = d-b$, or $a-c = d-b$. Without always repeating the principle, it is derived from observation, that its effect is to remove quantities from one side of an equation to another, changing their sign at the same time. But the beginner should not use this rule until he is perfectly familiar with the manner of using the principle. He should, until he has mastered a good many examples, continue the operation at full length, instead of using the rule, which is an abridgment of it. In fact it would be better, and not more prolix, to abandon the received phraseology, and in the example just cited, instead of saying "bring the term b to the other side of the equation," to say "subtract b from both sides," and instead of saying "bring c to the other side of the equation," to say "add c to both sides."

Suppose we have the fractions $\frac{3}{4}$, $\frac{1}{7}$, and $\frac{5}{14}$. If we multiply them all by the product of the denominators $4 \times 7 \times 14$, or 392, all the products will be whole numbers. They will be $\frac{3 \times 392}{4}$, $\frac{1 \times 392}{7}$, and $\frac{5 \times 392}{14}$, and since 392 is measured by 4, 3×392 is also measured by 4, and $\frac{3 \times 392}{4}$ is a whole number, and so on. But any common multiple of 4, 7, and 14 will serve as well. The least common multiple will therefore be the most convenient to use for this purpose. The least common multiple of 4, 7, and 14 is 28, and if the three fractions be multiplied by 28, the results will be whole numbers. The same also applies to algebraic fractions. Thus $\frac{a}{b}$, $\frac{c}{de}$, and $\frac{e}{bdf}$, will become simple expressions, if they are multiplied by $b \times de \times bdf$, or b^2d^2ef. But the most simple common multiple of b, de, and dbf, is $bdef$, which should be used in preference to b^2d^2ef.

This being premised, we can now reduce any equation which contains fractions to one which does not. For example, take the equation

$$\frac{x}{3} + \frac{2x}{5} = \frac{7}{10} - \frac{3 - 2x}{6}.$$

If we multiply both these equal quantities by any other,

the results will be equal. We choose, then, the least quantity, which will convert all the fractions into simple quantities, that is, the least common multiple of the denominators 3, 5, 10, and 6, which is 30. If we multiply both equal quantities by 30, the equation becomes

$$\frac{30x}{3} + \frac{60x}{5} = \frac{210}{10} - \frac{30(3-2x)}{6}. \qquad (1)$$

But $\frac{30x}{3}$ is $\frac{30}{3} \times x$ or $10x$, $\frac{60x}{5}$ is $\frac{60}{5} \times x$, or $12x$, etc.; so that we have

$$10x + 12x = 21 - 5(3-2x), \qquad (2)$$

$$\text{or} \quad 10x + 12x = 21 - (15 - 10x), \qquad (3)$$

$$\text{or} \quad 10x + 12x = 21 - 15 + 10x. \qquad (4)$$

Beginners very commonly mistake this process, and forget that the sign of subtraction, when it is written before a fraction, implies that the whole result of the fraction is to be subtracted from the rest. As long as the denominator remains, there is no need to signify this by putting the numerator between brackets, but when the denominator is taken away, unless this be done, the sign of subtraction belongs to the first term of the numerator only, and not to the whole expression. The way to avoid this mistake would be to place in brackets the numerators of all fractions which have the negative sign before them, and not to remove those brackets until the operation of subtraction has been performed, as is done in equation (4).

The following operations will afford exercise to the student, sufficient, perhaps, to enable him to avoid this error:

$$a + \frac{b-c+d-e}{f} = \frac{af+b-c+d-e}{f},$$

$$a - \frac{b-c+d-e}{f} = \frac{af-b+c-d+e}{f},$$

$$a+b+\frac{(a-b)^2}{a+b} = \frac{2a^2+2b^2}{a+b},$$

$$a+b-\frac{(a-b)^2}{a+b} = \frac{4ab}{a+b}.$$

We can now proceed with the solution of the equation. Taking up the equation (4) which we have deduced from it, subtract $10x$ from both sides, which gives $10x+12x-10x = 21-15$, or $12x = 6$: divide these equal quantities by 12, which gives $\frac{12x}{12} = \frac{6}{12}$, or $x = \frac{1}{2}$. This is the only value which x can have so as to make the given equation true, or, as it is called, to *satisfy* the equation. If instead of x we substitute $\frac{1}{2}$, we shall find that

$$\frac{\frac{1}{2}}{3} + \frac{2\times\frac{1}{2}}{5} = \frac{7}{10} - \frac{3-2\times\frac{1}{2}}{6}, \quad \text{or} \quad \frac{1}{6}+\frac{1}{5} = \frac{7}{10}-\frac{2}{6};$$

this we find to be true, since

$$\frac{1}{6}+\frac{1}{5} \text{ is } \frac{11}{30}, \quad \text{and} \quad \frac{7}{10}-\frac{2}{6} = \frac{22}{60}, \quad \text{and} \quad \frac{11}{30} = \frac{22}{60}.$$

In these equations of the first degree there is one unknown quantity and all the others are known. These known quantities may be represented by letters, and, as we have said, the first letters of the alphabet are commonly used for that purpose. We will now take an equation of exactly the same form as the last, putting letters in place of numbers:

$$\frac{x}{a} + \frac{bx}{c} = \frac{d}{c} - \frac{f - gx}{h}.$$

The solution of this equation is as follows: multiply both quantities by $aceh$, the most simple multiple of the denominators, it then becomes:

$$\begin{aligned} \frac{acehx}{a} + \frac{abcehx}{c} &= \frac{acdeh}{e} - \frac{aceh(f - gx)}{h}, \\ \text{or,}\quad cehx + abcehx &= acdh - ace(f - gx), \\ \text{or,}\quad cehx + abcehx &= acdh - acef + acegx. \end{aligned}$$

Subtract $acegx$ from both sides, and it becomes

$$\begin{aligned} cehx + abehx - acegx &= acdh - acef, \\ \text{or,}\quad (ceh + abeh - aceg)x &= acdh - acef. \end{aligned}$$

Divide both sides by $ceh + abeh - aceg$, which gives

$$x = \frac{acdh - acef}{ceh + abeh - aceg}.$$

The steps of the process in the second case are exactly the same as in the first; the same reasoning establishes them both, and the same errors, are to be avoided in each. If from this we wish to find the solution of the equation first given, we must substitute 3 for a, 2 for b, 5 for c, 7 for d, 10 for e, 3 for f, 2 for g, and 6 for h, which gives for the value of x,

$$\frac{3\times 5\times 7\times 6-3\times 5\times 10\times 3}{5\times 10\times 6+3\times 2\times 10\times 6-3\times 5\times 10\times 2},$$

$$\text{or,}\quad \frac{3\times 5\times 12}{3\times 2\times 10\times 6},\quad \text{or,}\quad \frac{180}{360},$$

which is $\frac{1}{2}$, the same as before.

If in one equation there are two unknown quantities, the condition is not sufficient to fix the values of the two quantities; it connects them, nevertheless, so that if one can be found the other can be found also. For example, the equation $x+y=8$ admits of an infinite number of solutions, for take x to represent any whole number or fraction less than 8, and let y represent what x wants of 8, and this equation is satisfied. If we have another equation of condition existing between the same quantities, for example, $3x-2y=4$; this second equation by itself has an infinite number of solutions: to find them, y may be taken at pleasure, and $x=\dfrac{4+2y}{3}$. Of all the solutions of the second equation, one only is a solution of the first; thus there is only one value of x and y which satisfies both the equations, and the finding of these values is the solution of the equations. But there are some particular cases in which every value of x and y which satisfies one of the equations satisfies the other also; this happens whenever one of the equations can be deduced from the other. For example, when $x + y = 8$, and $4x - 29 = 3 - 4y$, the second of these is the same as $4x + 4y = 3 + 29$, or $4x + 4y = 32$, which necessarily follows from the first equation.

If the solution of a problem should lead to two equations of this sort, it is a sign that the problem admits of an infinite number of solutions, or is what is called an indeterminate problem. The solution of equations of the first degree does not contain any peculiar difficulty; we shall therefore proceed to the consideration of the isolated negative sign.

CHAPTER IX.

ON THE NEGATIVE SIGN, ETC.

IF we wish to say that 8 is greater than 5 by the number 3, we write this equation $8 - 5 = 3$. Also to say that a exceeds b by c, we use the equation $a - b = c$. As long as some numbers whose value we know are subtracted from others equally known, there is no fear of our attempting to subtract the greater from the less; of our writing $3 - 8$, for example, instead of $8 - 3$. But in prosecuting investigations in which letters occur, we are liable, sometimes from inattention, sometimes from ignorance as to which is the greater of two quantities, or from misconception of some of the conditions of a problem, to reverse the quantities in a subtraction, for example to write $a - b$ where b is the greater of two quantities, instead of $b - a$. Had we done this with the sum of two quantities, it would have made no difference, because $a + b$ and $b + a$ are the same, but this is not the case with $a - b$ and $b - a$. For example, $8 - 3$ is easily understood; 3 can be taken from 8 and the remainder is 5; but $3 - 8$ is an impossibility, it requires you to take from 3 more than there is in 3, which is absurd. If such an expression as $3 - 8$ should be the answer to a problem, it would denote either that there was some absurdity inherent in the problem itself, or in the manner of putting it into an equation. Nevertheless, as such answers will occur, the student must be aware what sort of mistakes give rise to them, and in what manner they affect the process of investigation.

We would recommend to the beginner to make experience his only guide in forming his notions of these quantities, that is, to draw his rules from the observation of many results, not from any theory. The difficulties which encompass the theory of the negative sign are explained at best in a manner which would embarrass him: probably he would not see the difficulties themselves; too easy belief has always been the fault of young students in mathematics, and it is a great point gained to get them to start an objection. We shall observe the effect of this error in denoting a subtraction on every species of investigation to which we have hitherto come, and shall deduce rules which the student will recollect are the results of experience, not of abstract reasoning. The extensions to which he will be led have rendered Algebra much more general than it was before, have made it competent to the solution of many, very many questions which it could not have touched had they not been attended to. They do, in fact, constitute part of the groundwork of modern Algebra and should be considered by the student who is desirous of making his way into the depths of the science with the highest degree of attention. If he is well practised in the ordinary rules which have hitherto been explained, few difficulties can afterwards embarrass him, except those which arise from some confusion in the notions which he has formed upon this part of the subject.

For brevity's sake we hereafter use this phrase. Where the signs of every term in an expression are changed, it is said to have changed its form. Thus $+a - b$ and $+b - a$ are in different forms, and if a be greater than b, the first is the correct form and the second incorrect. An extension of a rule is made by which such a quantity as $3 - 8$ is written in a different way. Suppose that $+3 - 8$ is connected with any other number thus, $56 + 3 - 8$. This may be written $56 + 3 - (3 + 5)$, or $56 + 3 - 3 - 5$, or $56 - 5$. It appears, then, that $+3 - 8$, connected with any number is the same as -5 connected with that number; from this we say that $+3 - 8$, or $3 - 8$ is the same thing as -5, or $3 - 8 = -5$. This is another way of writing the equation $8 - 3 = 5$, and indicates equally that 8 is greater than 5 by 3. In the same way, $a - b = -c$ indicates that b is greater than a by the quantity c. If a be nothing, this equation becomes $-b = -c$, which indicates that $b = c$, since if the equation $a - b = -c$ be written in its true form $b - a = c$, and if $a = 0$, then $b = c$. We can now understand the following equations:

$$a - b + c - d = -e, \quad \text{or} \quad b + d - a - c = e,$$
$$2ab - a^2 - b^2 = -d - e, \quad \text{or} \quad a^2 + b^2 - 2ab = d + e.$$

We must not commence any operations upon such an equation as $a - b = -c$, until we have satisfied ourselves of the manner in which they should be performed, by reference to the correct form of the equation. This correct form is $b - a = c$. This gives $d + b - a = d + c$, or $d - (a - b) = d + c$. Write instead of $a - b$ its symbol $-c$, and then $d - (-c) = d + c$. Here we have performed an operation with $a - b$, which is no quantity, since a is less than b, but this is done because our present object is, in applying the common rules to such expressions, to watch the results and exhibit them in their real forms. The first side $d-(-c)$ is in a form in which we can attach no meaning to it, and the second side gives its real form $d+c$. The meaning of this expression is, that if with $a-b$, which we think to be a quantity, but which is not, since a is less than b, we follow the algebraical rule in subtracting $a-b$ from d, we shall thereby get the same result as if we had added the real quantity $b-a$ to d. If we make use of the form $d-(-c)$, it is because we can use it in such a manner as never to lose sight of its connexion with its real form $d+c$, and because we can establish rules which will lead us to the end of a process without any error, except those which we can correct as certainly at the end as at the beginning.

The rule by which we proceed, and which we shall establish by numerous examples, is, that wherever two like signs come together, the corresponding part of the real form has a positive sign, and wherever two unlike signs come together, the real form has a negative sign. Thus the real form of $d-(-c)$ is $d+c$. Again, take the real form $b-a=c$ of the equation $a-b=-c$, and it follows that $d-(b-a)=d-c$, or $d-b+a=d-c$, or $d+a-b=d-c$, or $d+(a-b)=d-c$. This is $d+(-c)=d-c$, another case in which the rule is verified. Again, multiply together $a-b$ and $m-n$, the product is $am-an-bm+bn$. This is the same product as arises from multiplying $b-a$ by $n-m$, written in a different order. If, then, $b - a = c$, and $n - m = p$, or $a - b = -c$, and $m - n = -p$, we find that $(-c) \times (-p) = cp$. By which result we mean that a mistake, in the form of both $a - b$ and $m - n$, will not produce a mistake in the form of their product, which remains what it would have been had the mistake not been made. Again

$$(n - m)(b - a) = bn - bm - an + am,$$
$$(n - m)(a - b) = an - am - bn + bm.$$

If the first product be real and equal to P, the second is represented by $-P$. The first is cp, the second is $(-c) \times p$, which gives

$$(-c) \times p = -cp.$$

That is, a mistake in the form of one factor only alters the form of the product. To distinguish the right form from the wrong one, we may prefix + to the first, and − to the second, and we may then recapitulate the results, and add others, which the student will now be able to verify.

The sign + placed before single quantities shows that the form of the quantity is correct; the sign − shows that it has been mistaken or changed.

$$a + (+b) = a + b, \qquad a + (-b) = a - b,$$
$$a - (+b) = a - b, \qquad a - (-b) = a + b,$$
$$(+a) \times (+b) = +ab, \qquad (+a) \times (-b) = -ab,$$
$$(-a) \times (-b) = +ab = (+a) \times (+b),$$

$$\frac{+a}{+b} = +\frac{a}{b},$$
$$\frac{+a}{-b} = -\frac{a}{b} = \frac{-a}{+b},$$
$$\frac{-a}{-b} = +\frac{a}{b},$$
$$\begin{aligned} -a \times -a &= +a^2, \\ -a \times -a \times -a &= +a^2 \times -a = -a^3, \\ -a \times -a \times -a \times -a &= -a^3 \times -a = +a^4, \\ \text{etc.,} \quad & \quad \text{etc.} \end{aligned}$$

We see, then, that a change in the form of any quantity changes the form of those powers whose exponent is an odd number, but not of those whose exponent is an even number. By these rules we shall be able to tell what changes would be made in an expression by altering the forms of any of its letters. It may be fairly asked whether we are not changing the meaning of the signs + and −, in making $+a$ stand for an expression in which we do not alter the signs, and $-a$ for one in which the signs are altered. The change is only in name, for since the rule of addition is, "annex the expressions which are to be added without altering the signs of either," or "annex the expressions without altering the form of either;" the quantity $a + b$, which is the sum of the two expressions a and b, stands for the same as $+a + b$, in which the new notion of the sign + is used, viz., the expressions a and b are annexed with unaltered forms, which is denoted by writing together $+a$ and $+b$. Again, the rule for subtraction is, "change the sign of the subtrahend or expression which is to be subtracted, and annex the result to the other expression," or "change the form of the subtrahend and annex it to the other," which, the expressions being a and b, is written $a - b$, which answers equally well to the second notion of the sign $-$, since $+a - b$ indicates that a and b are to be annexed, the first without, the second with a change of form. These ideas of the signs $+$ and $-$ give, therefore, in practice, the same results as the former ones, and, in future, the two meanings may be used indiscriminately. But when a single term is used, such as $+a$ or $-a$, the last acquired notions of $+$ and $-$ are always understood.

This much being premised, we can see, by numberless instances, that, if the form of a quantity is to be changed, it matters nothing whether it is changed at the beginning of the process, or whether we wait till the end, and then follow the rules above mentioned. This is evident to the more advanced student, from the nature of the rules themselves, but the beginner should satisfy himself of this fact from experience. We now give a proof of this, as far as one expression can prove it, in the solution of the equations,

$$\frac{a^2}{b} + ax = \frac{a^2x}{b} + a - b,$$

$$\text{and} \quad \frac{a^2}{b} - ax = \frac{a^2x}{b} - a - b,$$

which two equations only differ in the form in which a appears. For, if the form of a in the first equation be altered, that of $\frac{a^2}{b}$ and $\frac{a^2x}{b}$ is unaltered, $+ax$ becomes $-ax$, and $+a$ becomes $-a$. We now solve the two equations in opposite columns.

$$\begin{aligned} \frac{a^2}{b} + ax &= \frac{a^2x}{b} + a - b, \\ a^2 + abx &= a^2x + ab - b^2, \\ a^2 - ab + b^2 &= a^2x - abx \\ &= (a^2 - ab)x, \\ x &= \frac{a^2 - ab + b^2}{a^2 - ab}; \end{aligned} \qquad \begin{aligned} \frac{a^2}{b} - ax &= \frac{a^2x}{b} - a - b, \\ a^2 - abx &= a^2x - ab - b^2, \\ a^2 + ab + b^2 &= a^2x + abx \\ &= (a^2 + ab)x, \\ x &= \frac{a^2 + ab + b^2}{a^2 + ab}. \end{aligned}$$

The only difference between these expressions arises from the different form of a in the two. If, in either of them, $-a$ be put instead of $+a$, and the rules laid down be followed, the other will be produced. We see, then, that a simple alteration of the form of a in the original equation produces no other change in the result, or in any one of the steps which lead to that result, except a simple alteration in the form of a. From this it follows that, having the solution of an equation, we have also the solution of all the equations which can be formed from it, by altering the form of the different known quantities which are contained in it. And, as all problems can be reduced to equations, the solution of one problem will lead us to the solution of others, which differ from the first in producing equations in which some of the known quantities are in different forms. Also, in every identical equation, the form of one or more of its quantities may be altered throughout, and the equation will still remain identically true. For example,

$$\frac{a^3 - b^3}{a - b} = a^2 + ab + b^2.$$

Change $+b$ into $-b$, and this equation will become

$$\frac{a^3 + b^3}{a + b} = a^2 - ab + b^2,$$

which last, common division will show to be true.

Again, suppose than when a, b, and c are in a given form, which we denote by $+a$, $+b$, and $+c$, the solution of a problem is,

$$x = \frac{b^2 - 4ac}{a + c - b}.$$

The following table will show the alterations which take place in x when the forms of a, b, and c are changed in different manners, and the verification of it will be an exercise for the student.

FORMS OF a, b, AND c.	VALUES OF x.
$+a,\ +b,\ +c$	$\frac{b^2 - 4ac}{a + c - b},$
$+a,\ +b,\ -c$	$\frac{b^2 + 4ac}{a - c - b},$
$+a,\ -b,\ -c$	$\frac{b^2 + 4ac}{a - c + b},$
$-a,\ +b,\ -c$	$-\frac{b^2 - 4ac}{b + a + c},$
$-a,\ -b,\ -c$	$\frac{b^2 - 4ac}{b - a - c}.$

Also, the expression for x may be written in the following different ways, the forms of a, b, and c remaining the same:

$$\frac{b^2 - 4ac}{a + c - b}, \quad -\frac{b^2 - 4ac}{b - a - c}, \quad -\frac{4ac - b^2}{a + c - b}, \quad \frac{4ac - b^2}{b - a - c}.$$

We now proceed to apply these principles to the solution of the following problems:

Two couriers, A and B, in the course of a journey between the towns C and D, are at the same moment of time at A and B. A goes m miles, and B, n miles an hour. At what point between C and D are they together? It is evident that the answer depends upon whether they are going in the same or opposite directions, whether A goes faster or slower than B, and so on. But all these, as we shall see, are included in the same general problem, the difference between them corresponding to the different forms of the letters which we shall have occasion to use. After solving the different cases which present themselves, each upon its own principle, we shall compare the results in order to establish their connexion. Let the distance AB be called a.

Case first.—Suppose that they are going in the same direction from C to D, and that m is greater than n. They will then meet at some point between B and D. Let that point be H, and let AH be called x. Then A travels through AH, or x, in the time during which B travels through BH, or $x-a$. But, since A goes m miles an hour, he travels the distance x in $\frac{x}{m}$ hours. Again, B travels the distance $x-a$ in $\frac{x-a}{n}$ hours. These times are the same, and, therefore,

$$\frac{x}{m}=\frac{x-a}{n}, \quad \text{or} \quad x=\frac{ma}{m-n}=AH$$

$$\text{and} \quad x-a=\frac{na}{m-n}=BH.$$

The time which elapses before they meet is

$$\frac{x}{m} \quad \text{or} \quad \frac{a}{m-n}.$$

Case second.—Suppose them now moving in the same direction as before, but let B move faster than A. They never will meet after they come to A and B, since B is continually gaining upon A, but they must have met at some point before reaching A and B. Let that point be H, and, as before, let $AH=x$. Then since A travels through HA or x in the time

during which B travels through HB, or $x+a$, in the same manner as in the last case, we show that

$$\frac{x}{m} = \frac{x+a}{n}, \quad \text{or} \quad x = \frac{ma}{n-m} = AH$$
$$\text{and} \quad x + a = \frac{na}{n-m} = BH.$$

The time elapsed is $\dfrac{a}{n-m}$.

Case third.—If they are moving from D to C, and if B moves faster than A, the point H is the same as in the last case, since, if having in the last case arrived at A and B, they move back again at the same rate, they will both arrive at the point H together. The answers in this case are therefore the same as in the last.

Case fourth.—Similarly, if they are moving from D to C, and A moves faster than B, the answers are the same as in the first case, since this is a reverse of the first case, as the third is of the second. We reserve for the present the case in which they move equally fast, as another species of difficulty is involved which has no connexion with the present subject. We shall return to it hereafter.

Case fifth.—Suppose them now moving in contrary directions, viz.: A towards D and B towards C. Whether A moves faster or slower than B, they must now meet somewhere between A and B; as before let them meet in H, and let $AH = x$. Then A moves through AH, or x, in the same

time as B moves through BH, or $a - x$. Therefore

$$\frac{x}{m} = \frac{a-x}{n}, \quad \text{or}$$

$$x = \frac{ma}{m+n},$$

$$a - x = \frac{na}{m+n}.$$

The time elapsed is $\dfrac{a}{m+n}$.

Case sixth.—Let them be moving in contrary directions, but let A be moving towards C, and B towards D. They will then have met somewhere between A and B, and as this is only the reverse of the last case, just as the fourth is of the first, or the third of the second, the answers are the same. We now exhibit the results of these different cases in a table, stating the circumstances of each case, and also whether the time of meeting is before or after the instant which finds them at A and B.

Now $\dfrac{a}{m-n}$ and $\dfrac{a}{n-m}$ are the same quantity written in different forms, for $n - m$ is $-(m - n)$; and according to the rules

$$\frac{a}{n-m} = -\frac{a}{m-n}.$$

Similarly

$$\frac{ma}{n-m} = -\frac{ma}{m-n},$$

Circumstances of the case.	Direction of the point H.	Value of AH.	Value of BH.	Time of meeting
1. Both move from C to D, A moves faster than B.	Between B and D.	$\frac{ma}{m-n}$	$\frac{na}{m-n}$	$\frac{a}{m-n}$ after.
2. Both move from C to D, A moves slower than B.	Between A and C.	$\frac{ma}{n-m}$	$\frac{na}{n-m}$	$\frac{a}{n-m}$ before.
3. Both move from D to C, A moves slower than B.	Between A and C.	$\frac{ma}{n-m}$	$\frac{na}{n-m}$	$\frac{a}{n-m}$ after.
4. Both move from D to C, A moves faster than B.	Between B and D.	$\frac{ma}{m-n}$	$\frac{na}{m-n}$	$\frac{a}{m-n}$ before.
5. A moves towards D and B towards C.	Between A and B.	$\frac{ma}{m+n}$	$\frac{na}{m+n}$	$\frac{a}{m+n}$ after.
6. A moves towards C and B towards D.	Between A and B.	$\frac{ma}{m+n}$	$\frac{na}{m+n}$	$\frac{a}{m+n}$ before.

and so on.

We see also, that in the first and second cases, which differ in this, that AH falls to the right in the first, and to the left in the second, the forms of AH are different, there being $\frac{ma}{m-n}$ in the first, and $-\frac{ma}{m-n}$ in the second. Again, in the same cases, in the first of which the time of meeting is *after*, and in the second *before* the moment of being at A and B, we see a difference of form in the value of that time; in the first it is $\frac{a}{m-n}$, and in the second $-\frac{a}{m-n}$, or $\frac{a}{n-m}$. The same remarks apply to the third and fourth examples. Again, in the first and fifth cases, which only differ in this, that B is moving towards D in the first, and in the contrary direction towards C in the fifth, the values of AH, and of the time, may be deduced from the first by changing the form of n, and writing $+n$, instead of $-n$. The expression for BH in the first, if the form of n be likewise changed, becomes $-\frac{na}{m+n}$, which is the value of BH in the fifth, but in a different form. But we observe that BH falls to the left of B in the fifth, whereas it fell to the right in the first. Again, in the first and sixth examples, which differ in this that A moves towards D in the first and towards C in the sixth, the value of AH in the sixth may be deduced from that of AH in the first by changing the form of m, which change makes AH become $\frac{-ma}{-m-n}$, or $\frac{-ma}{-(m+n)}$, or $\frac{ma}{m+n}$. If we alter the value of the time in the first, in the same manner, it becomes $\frac{a}{-m-n}$, or $-\frac{a}{m+n}$, which is of a different form from that in the sixth; but it must also be observed that the first is after and the other before the moment when they are at A and B. In the fifth and sixth examples which differ in this, that the direction in which both are going is changed, since in the fifth they move towards one another, and in the sixth away from one another, the values of AH and BH in the one may be deduced from those in the other by a change of form, both in m and n, which gives the same values as before. But if m and n change their forms in the expression for the time, the value in the sixth case is $\frac{a}{-m-n}$, or $-\frac{a}{m+n}$. Also the time in the fifth case is after the moment at which they are

Circumstances of the case.	Direction of the point H.	Value of AH.	Value of BH.	Time of meeting
1. Both move from C to D, A moves faster than B.	Between B and D.	$\frac{ma}{m-n}$	$\frac{na}{m-n}$	$\frac{a}{m-n}$ after.
2. Both move from C to D, A moves slower than B.	Between A and C.	$\frac{ma}{n-m}$	$\frac{na}{n-m}$	$\frac{a}{n-m}$ before.
3. Both move from D to C, A moves slower than B.	Between A and C.	$\frac{ma}{n-m}$	$\frac{na}{n-m}$	$\frac{a}{n-m}$ after.
4. Both move from D to C, A moves faster than B.	Between B and D.	$\frac{ma}{m-n}$	$\frac{na}{m-n}$	$\frac{a}{m-n}$ before.
5. A moves towards D and B towards C.	Between A and B.	$\frac{ma}{m+n}$	$\frac{na}{m+n}$	$\frac{a}{m+n}$ after.
6. A moves towards C and B towards D.	Between A and B.	$\frac{ma}{m+n}$	$\frac{na}{m+n}$	$\frac{a}{m+n}$ before.

(Table of Page 115 Repeated.)

at A and B, and in the sixth case it is before. From these comparisons we deduce the following general conclusions:

1. If we take the first case as a standard, we may, from the values which it gives, deduce those which hold good in all the other cases. If a second case be taken, and it is required to deduce answers to the second case from those of the first, this is done by changing the sign of all those quantities whose directions are opposite in the second case to what they are in the first, and if any answer should appear in a negative form, such as $\frac{ma}{m-n}$, when m is less than n, which may be written thus $-\frac{ma}{n-m}$, it is a sign that the quantity which

it represents is different in direction in the first and second cases. If it be a right line measured from a given point in all the cases, such as AH, it is a sign that AH falls on the left in the second case, if it fell on the right in the first case, and the converse. If it be the time elapsed between the moment in which the couriers are at A and B and their meeting, it is a sign that the moment of meeting is before the other, in the second case, if it were after it in the first, and the converse. We see, then, that these six cases can be all contained in one if we apply this rule, and it is indifferent which of the cases is taken as the standard, provided the corresponding alterations are made to determine answers to the rest.

This detail has been entered into in order that the student may establish from his own experience the general principle which will conclude this part of the subject. Further illustration is contained in the following problem:

A workman receives a shillings a day for his labor or a proportion of a shillings for any part of a day which he works. His expenses are b shillings every day, whether he works or no, and after m days he finds that he has gained c shillings. How many days did he work? Let x be that number of days, x being either whole or fractional; then for his work he receives ax shillings, and during the m days his expenditure is bm shillings, and since his gain is the difference between his receipts and expenditure:

$$ax - bm = c,$$

$$\text{or} \quad x = \frac{bm + c}{a}.$$

Now suppose that he had worked so little as to lose c shillings instead of gaining anything. The equation from which x is derived is now

$$bm - ax = c,$$

which, when its form is changed, becomes

$$ax - bm = -c,$$

an equation which only differs from the former in having $-c$ written instead of c. The solution of the equation is

$$x = \frac{bm - c}{a},$$

which only differs from the former in having $-c$ instead of $+c$. It appears then that we may alter the solution of a problem which proceeds upon the supposition of a gain into the solution of one which supposes an equal loss, by changing the form of the expression which represents that gain; and also that if the answer to a problem which we have solved upon the supposition of a gain should happen to be negative, suppose it $-c$, we should have proceeded upon the supposition that there is a loss and should in that case have found a loss, c. When such principles as these have been established, we have no occasion to correct an erroneous solution by recommencing the whole process, but we may, by means of the form of the answer, set the matter right at the end. The principle is, that a negative solution indicates that the nature of the answer is the very reverse of that which it was supposed to be in the solution; for example, if the solution supposes a line measured in feet in one direction, a negative answer, such as $-c$, indicates that c feet must be measured in the opposite direction; if the answer was thought to be a number of days *after* a certain epoch, the solution shows that it is c days before that epoch; if we supposed that A was to receive a certain number of pounds, it denotes that he is to pay c pounds, and so on, In deducing this principle we have not made any supposition as to what $-c$ is; we have not asserted that it indicates the subtraction of c from 0; we have derived the result from observation only, which taught us first to deduce rules for making that alteration in the result which arises from altering $+c$ into $-c$ at the commencement; and secondly, how to make the solution of one case of a problem serve to determine those of all the others. By observation then the student must acquire his conviction of the truth of these rules, reserving all metaphysical discussion upon such quantities as $+c$ and $-c$ to a later stage, when he will be better prepared to understand the difficulties of the subject. We now proceed to another class of difficulties, which are generally, if possible, as much misconceived by the beginner as the use of the negative sign.

Take any fraction $\frac{a}{b}$. Suppose its numerator to remain the same, but its denominator to decrease, by which means the fraction itself is increased. For example, $\frac{5}{12}$ is greater than $\frac{5}{20}$ or the twelfth part of 5 is greater than its twentieth part. Similarly, $\frac{2\frac{1}{2}}{4\frac{1}{6}}$ is greater than $\frac{2\frac{1}{2}}{\frac{27}{2}}$, etc. If, then, b be diminished more and more, the fraction $\frac{a}{b}$ becomes greater and greater, and there is no limit to its possible increase. To show this, suppose that b is a part of a, or that $b = \frac{a}{m}$. Then $\frac{a}{b}$ or $\frac{a}{\frac{a}{m}}$ is m. Now since b may diminish so as to be equal to any part of a, however small, that is, so as to make m any number, however great, $\frac{a}{b}$ which is $= m$ may be any number however great. This diminution of b, and the consequent increase of $\frac{a}{b}$, may be carried on to any extent, which we may state in these words: As the quantity b becomes nearer and nearer to 0, the fraction $\frac{a}{b}$ increases, and in the interval in which b passes from its first magnitude to 0, the fraction $\frac{a}{b}$ passes from its first value through every possible greater number. Now, suppose that the solution of a problem in its most general form is $\frac{a}{b}$, but that in one particular case of that problem b is $= 0$. We have then instead of a solution $\frac{a}{0}$, a symbol to which we have not hitherto given a meaning.

To take an instance: return to the problem of the two couriers, and suppose that they move in the same direction from C to D (*Case first*) at the same rate, or that $m = n$. We find that $AH = \frac{ma}{m-n}$ or $\frac{ma}{n-m}$ or $\frac{ma}{0}$. On looking at the equation which produced this result we find that it becomes $\frac{x}{m} = \frac{x-a}{m}$, or $x = x - a$, which is impossible. On looking at the manner in which this equation was formed, we find that it was made on the supposition that A and B are together at some point, which in this case is also impossible, since if they move at the same rate, the same distance which separated them at one moment will separate them at any other, and they will never be together, nor will they ever have been together on the other side of A. The conclusion to be drawn is, that such an equation as $x = \frac{a}{0}$ indicates that the supposition from which x was deduced can never hold good. Nevertheless in the common language of algebra it is said that they meet at an infinite distance, and that $\frac{a}{0}$ is infinite. This phrase is one which in its literal meaning is an absurdity, since there is no such thing as an infinite number, that is a number which is greater than any other, because the mind can set no bounds to the magnitude of the numbers which it can conceive, and whatever number it can imagine, however great, it can imagine the next to it. But as the use of the phrase is very general, the only method is to attach a meaning which shall not involve absurdity or confusion of ideas. The phrase used is this: When $c = b$, $\frac{a}{c-b} = \frac{a}{0}$ and is infinitely great. The student should always recollect that this is an abbreviation of the following sentence. "The fraction $\frac{a}{c-b}$ becomes greater and greater as c approaches more and more near to b; and if c, setting out from a certain value, should change gradually until it becomes equal to b, the fraction $\frac{a}{c-b}$ setting out also from a certain value, will attain any magnitude however great, before c becomes equal to b." That is, before a fraction can assume the form $\frac{a}{0}$, it must increase without limit. The symbol ∞ is used to denote such a fraction, or in general any quantity which increases without limit. The following equation will tend to elucidate the use of this symbol. In the problem of the two couriers, the equation which gave the result $\frac{ma}{0}$ was $\frac{x}{m} = \frac{x-a}{m}$, or $x = x - a$, which is evidently impossible. Nevertheless, the larger x is taken the more near is this equation to the truth, as may be proved by dividing both sides by x, when it becomes $1 = 1 - \frac{a}{x}$, which is never exactly true. But the fraction $\frac{a}{x}$ decreases as x increases, and by taking x sufficiently great may be reduced to any degree of smallness. For example, if it is required that $\frac{a}{x}$ should be as small as $\frac{1}{10000000}$ of a unit, take x as great as $10000000a$, and the fraction becomes $\frac{a}{10000000a}$, or $\frac{1}{10000000}$. But as

$\frac{a}{x}$ becomes smaller and smaller, the equation $1 = 1 - \frac{a}{x}$ becomes nearer and nearer the truth, which is expressed by saying that when $1 = 1 - \frac{a}{x}$, or $x = x - a$, the solution is $x = \infty$. In the solution of the problem of the two couriers this does not appear to hold good, since when $m = n$ and $x = \frac{ma}{0}$ the same distance a always separates them, and no travelling will bring them nearer together. To show what is meant by saying that the greater x is, the nearer will it be a solution of the problem, suppose them to have travelled at the same rate to a great distance from C. They can never

come together unless CA becomes equal to CB, or A coincides with B, which never happens, since the distance AB is always the same. But if we suppose that they have met, though an error always will arise from this false supposition, it will become less and less as they travel farther and farther from C. For example, let $CA = 10000000AB$, then the supposing that they have met, or that B and A coincide, or that $BA = 0$, is an error which involves no more than $\frac{1}{10000000}$ of AC; and though AB is always of the same numerical magnitude, it grows smaller and smaller in comparison with AC, as the latter grows greater and greater.

Let us suppose now that in the problem of the two couriers they move in the same direction at the same rate, as in the case we have just considered, but that moreover they set out from the same point, that is, let $a = 0$. It is now evident that they will always be together, that is, that any value of x whatever is an answer to the question. On looking at the value of AH, or $\frac{ma}{m-n}$, we find the numerator and denominator both equal to 0, and the value of AH appears in the form $\frac{0}{0}$. But from the problem we have found that one value cannot be assigned to AH, since every point of their course is a point where they are together. The solution of the following equation will further elucidate this. Let

$$ax + by = c,$$
$$dx + ey = f,$$

from which, by the common method of solution, we find

$$x = \frac{ce - bf}{ae - bd}, \qquad y = \frac{af - cd}{ae - bd}.$$

Now, let us suppose that $ce = bf$ and $ae = bd$. Dividing the first of these by the second, we find

$$\frac{ce}{ae} = \frac{bf}{bd}, \quad \text{or} \quad \frac{c}{a} = \frac{f}{d}, \quad \text{or} \quad cd = af.$$

The values both of x and y in this case assume the form $\frac{0}{0}$; to find the cause of this we must return to the equations. If we divide the first of these by c, and the second by f, we find that

$$\frac{a}{c}x + \frac{b}{c}y = 1,$$
$$\frac{d}{f}x + \frac{e}{f}y = 1.$$

But the equations $ce = bf$ and $cd = af$ give us $\frac{b}{c} = \frac{e}{f}$ and $\frac{a}{c} = \frac{d}{f}$, that is, these two are, in fact, one and the same equation repeated, from which, as has been explained before, an infinite number of values of x and y can be found; in fact, any value may be given to x provided y be then found from the equation. We see that in these instances, when the value of any quantity appears in the form $\frac{0}{0}$, that quantity admits of an infinite number of values, and this indicates that the conditions given to determine that quantity are not sufficient. But this is not the only cause of the appearance of a fraction in the form $\frac{0}{0}$. Take the identical equation

$$\frac{a^2 - b^2}{c(a - b)} = \frac{a + b}{c}.$$

When a approaches towards b, $a + b$ approaches towards $2b$, and $a^2 - b^2$ and $a - b$ approach more and more nearly towards 0. If $a = b$ the equation assumes this form:

$$\frac{0}{0} = \frac{2b}{c}.$$

This may be explained thus: if we multiply the numerator and denominator of the fraction $\frac{A}{B}$ by $a - b$ (which does not alter its value) it becomes $\frac{Aa - Ab}{Ba - Bb}$. If in the course of an investigation this has been done when the two quantities a and b are equal to one another, the fraction $\frac{A}{B}$ or $\frac{Aa - Ab}{Ba - Bb}$ will appear in the form $\frac{0}{0}$. But since the result would have been $\frac{A}{B}$ had that multiplication not been performed, this last fraction must be used instead of the unmeaning form $\frac{0}{0}$. Thus the fraction $\frac{a^2 - b^2}{c(a - b)}$ or $\frac{(a + b)(a - b)}{c(a - b)}$ is the fraction $\frac{a + b}{c}$ after its numerator and denominator have been multiplied by $a - b$, and may be used in all cases except that in which $a = b$. When the form $\frac{0}{0}$ occurs, the problem must be carefully examined in order to ascertain the reason.

CHAPTER X.

EQUATIONS OF THE SECOND DEGREE.

EVERY operation of algebra is connected with another which is exactly opposite to it in its effects. Thus addition and subtraction, multiplication and division, are reverse operations, that is, what is done by the one is undone by the other. Thus $a + b - b$ is a, and $\frac{ab}{b}$ is a. Now in connexion with the raising of powers is a contrary operation called the extraction of roots. The term *root* is thus explained: We have seen that aa, or a^2, is called the square of a; from which a is called the square root of a^2. As 169 is called the square of 13, 13 is called the square root of 169. The following table will show how this phraseology is carried on.

a is called the square root of a^2,				denoted by		$\sqrt{a^2}$,
a "	"	"	cube root of a^3,	"	"	$\sqrt[3]{a^3}$,
a "	"	"	fourth root of a^4,	"	"	$\sqrt[4]{a^4}$,
a "	"	"	fifth root of a^5,	"	"	$\sqrt[5]{a^5}$,
etc.,			etc.,			etc.

If b stand for a^5, $\sqrt[5]{b}$ stands for a, and the foregoing table may be represented thus:

$$\text{If } a^2 = b, \quad a = \sqrt{b},$$
$$\text{if } a^3 = b, \quad a = \sqrt[3]{b}, \quad \text{etc.}$$

The usual method of proceeding is to teach the student to extract the square root of any algebraical quantity immediately after the solution of equations of the first degree. We would rather recommend him to omit this rule until he is acquainted with the solution of equations of the second degree, except in the cases to which we now proceed. In arithmetic, it must be observed that there are comparatively very few numbers of which the square root can be extracted. For example, 7 is not made by the multiplication either of any whole number or fraction by itself. The first is evident; the second cannot be readily proved to the beginner, but he may, by taking a number of instances, satisfy himself of this, that no fraction which is really such, that is whose numerator is not measured by its denominator, will give a whole number when multiplied by itself, thus $\frac{4}{2} \times \frac{4}{3}$ or $\frac{16}{9}$ is not a whole number, and so on. The number 7, therefore, is neither the square of a whole number, nor of a fraction, and, properly speaking, has no square root. Nevertheless, fractions can be found extremely near to 7, which have square roots, and this degree of nearness may be carried to any extent we please. Thus, if required, between 7 and $7\frac{1}{1000000000}$ could be found a fraction which has a square root, and the fraction in the last might be decreased to any extent whatever, so that though we cannot find a fraction whose square is 7, we may nevertheless find one whose square is as near to 7 as we please. To take another example, if we multiply 1.4142 by itself the product is 1.99996164, which only differs from 2 by the very small fraction .00003836, so that the square of 1.4142 is very nearly 2, and fractions might be found whose squares are still nearer to 2. Let us now suppose the following problem. A man buys a certain number of yards of stuff for two shillings, and the number of yards which he gets is exactly the number of shillings which he gives for a yard. How many yards does he buy? Let x be this number, then $\frac{2}{x}$ is the price of one yard, and $x = \frac{2}{x}$ or $x^2 = 2$. This, from what we have said, is impossible, that is, there is no exact number of yards, or parts of yards, which will satisfy the conditions; nevertheless, 1.4142 yards will nearly do it, 1.4142136 still more nearly, and if the problem were ever proposed in practice, there would be no difficulty in solving it with sufficient nearness for any purpose. A problem, therefore, whose solution contains a square root which cannot be extracted, maybe rendered useful by approximation to the square root.

Equations of the second degree, commonly called quadratic equations, are those in which there is the second power, or square of an unknown quantity: such as $x^2 - 3 = 4x^2 - 15$, $x^2 + 3x = 2x^2 - x - 1$, etc. By transposition of their terms, they may always be reduced to one of the following forms:

$$ax^2 + b = 0,$$
$$ax^2 - b = 0,$$
$$ax^2 + bx + c = 0,$$
$$ax^2 - bx + c = 0,$$

$$ax^2 + bx - c = 0,$$
$$ax^2 - bx - c = 0.$$

For example, the two equations given above, are equivalent to $3x^2 - 12 = 0$, and $x^2 - 4x - 1 = 0$, which agree in form with the second and last. In order to proceed to each of these equations, first take the equation $x^2 = a^2$. This equation is the same as $x^2 - a^2 = 0$, or $(x + a)(x - a) = 0$. Now, in order that the product of two or more quantities may be equal to nothing, it is sufficient that *one* of those quantities be nothing, and therefore a value of x may be derived from either of the following equations:

$$x - a = 0,$$
$$\text{or}\quad x + a = 0,$$

the first of which gives $x = a$, and the second $x = -a$. To elucidate this, find x from the following equation:

$$(3x + a)(a^3 + x^3) = (x^2 + ax)(a^2 + ax + 2x^2);$$

develop this equation, and transpose all its terms on one side, when it becomes

$$x^4 - 2a^2x^2 + a^4 + 2a^3x - 2ax^3 = 0,$$
$$\text{or}\quad (x^2 - a^2)^2 - 2ax(x^2 - a^2) = 0,$$
$$\text{or}\quad (x^2 - a^2)(x^2 - 2ax - a^2) = 0.$$

This last equation is true when $x^2 - a^2 = 0$, or when $x^2 = a^2$, which is true either when $x = +a$, or $x = -a$. If in the original equation $+a$ is substituted instead of x, the result is $4a \times 2a^3 = 2a^2 \times 4a^2$; if $-a$ be substituted instead of x, the result is $0 = 0$, which show that $+a$ and $-a$ are both correct values of x. We have here noticed, for the first time, an equation of condition which is capable of being solved by more than one value of x. We have found two, and shall find more when we can solve the equation $x^2 - 2ax - a^2 = 0$, or $x^2 - 2ax = a^2$. Every equation of the second degree, if it has one value of x, has a second, of which $x^2 = a^2$ is an instance, where $x = \pm a$, in which by the double sign $\pm$ is meant, that either of them may be used at pleasure. We now proceed to the solution of $ax^2 - bx + c = 0$. In order to understand the nature of this equation, let us suppose that we take for x such a value, that $ax^2 - bx + c$, instead of being equal to 0, is equal to y, that is

$$y = ax^2 - bx + c,^* \tag{1}$$

in which the value of y depends upon the value given to x, and changes when x changes. Let m be one of those quantities which, when substituted instead of x, makes $ax^2 - bx + c$ equal to nothing, in which case m is called a root of the equation,

$$ax^2 - bx + c = 0, \tag{2}$$

*In the investigations which follow, a, b, and c are considered as having the sign which is marked before them, and no change of form is supposed to take place.

and it follows that

$$am^2 - bm + c = 0. \tag{3}$$

Subtract (3) from (1), the result of which is

$$y = a(x^2 - m^2) - b(x - m) = (x - m)(a\,\overline{x + m} - b).$$

Here y is evidently equal to 0, when $x = m$, as we might expect from the supposition which we made; but it is also nothing when $a(x + m) - b = 0$; there is, therefore, another value of x, for which $y = 0$; if we call this n we find it from the equation $a(n + m) - b = 0$,

$$\text{or} \quad n + m = \frac{b}{a}. \tag{4}$$

In (3) substitute for b its value derived from (4), from which $b = a(n + m)$; it then becomes

$$am^2 - am(n + m) + c = 0, \quad \text{or} \quad c - amn = 0,$$

which gives

$$mn = \frac{c}{a}. \tag{5}$$

Substitute in (1) the values of b and c derived from (4) and (5), which gives

$$\begin{aligned} y &= ax^2 - a(m + n)x + amn \\ &= a(x^2 - \overline{m + n}\,x + mn). \end{aligned}$$

Now the second factor of this expression arises from multiplying together $\overline{x-m}$ and $\overline{x-n}$; therefore,

$$y = a(x-m)(x-n). \tag{6}$$

To take an example of this, let $y = 4x^2 - 5x + 1$. Here when $x = 1$, $y = 4 - 5 + 1 = 0$, and therefore $m = 1$. If we divide $4x^2 - 5x + 1$ by $x - 1$, the quotient (which is without remainder) is $4x - 1$, and therefore

$$y = (x-1)(4x-1).$$

This is also nothing when $4x - 1 = 0$, or when x is $\frac{1}{4}$. Therefore $n = \frac{1}{4}$, and $y = 4(x-1)(x-\frac{1}{4})$, a result coinciding with that of (6). If, therefore, we can find one of the values of x which satisfy the equation $ax^2 - bx + c = 0$, we can find the other and can divide $ax^2 - bx + c$ into the factors a, $x - m$ and $x - n$, or

$$ax^2 - bx + c = a(x-m)(x-n).$$

If we multiply $x + m$ by $x + n$, the only difference between $(x+m)(x+n)$ and $(x-m)(x-n)$ is in the sign of the term which contains the first power of x. If, therefore,

$$ax^2 - bx + c = a(x-m)(x-n),$$

it follows that

$$ax^2 + bx + c = a(x+m)(x+n).$$

We now take the expression $ax^2 - bx - c$. If there is one value of x which will make this quantity equal to 0, let this be m, and let

$$y = ax^2 - bx - c.$$

Then

$$0 = am^2 - bm - c,$$

from which

$$\begin{aligned} y &= a(x^2 - m^2) - b(x - m) \\ &= (x - m)(a\overline{x + m} - b) \\ &= (x - m)(ax + am - b). \end{aligned}$$

Let $\dfrac{am - b}{a}$ be called n, or let $am - b = an$; then

$$\begin{aligned} y &= (x - m)(ax + an) \\ &= a(x - m)(x + n). \end{aligned}$$

As an example, it may be shown that

$$3x^2 - x - 2 = 3(x - 1)(x + \tfrac{2}{3}).$$

Again, with regard to $ax^2 + bx - c$, since $(x + m)(x - n)$ only differs from $(x - m)(x + n))$ in the sign of the term which contains the first power of x, it is evident that

$$\begin{aligned} \text{if} \quad ax^2 - bx - c &= a(x - m)(x + n), \\ ax^2 + bx - c &= a(x + m)(x - n). \end{aligned}$$

Results similar to those of the first case may be obtained for all the others, and these results may be arranged in the following way. In the first and third, m is a quantity, which, when substituted for x, makes $y = 0$, and in the second and fourth m and n are the same as in the first and third.

1st $$y = ax^2 - bx + c = a(x - m)(x - n),$$

$$m + n = \frac{b}{a}, \qquad mn = \frac{c}{a}.$$

2d $$y = ax^2 + bx + c = a(x + m)(x + n),$$

$$m + n = \frac{b}{a}, \qquad mn = \frac{c}{a}.$$

3d $$y = ax^2 - bx - c = a(x - m)(x + n),$$

$$m - n = \frac{b}{a}, \qquad mn = \frac{c}{a}.$$

4th $$y = ax^2 + bx - c = a(x + m)(x - n),$$

$$m - n = \frac{b}{a}, \qquad mn = \frac{c}{a}.$$

We must now inquire in what cases a value can be found for x, which will make $y = 0$ in these different expressions, and in this consists the solution of equations of the second degree.

Let

$$y = ax^2 - bx + c, \tag{1}$$

and observe that $(2ax - b)^2 = 4a^2x^2 - 4abx + b^2$. Multiply both sides of (1) by $4a$, which gives

$$4ay = 4a^2x^2 - 4abx + 4ac. \tag{2}$$

Add b^2 to the first two terms of the second side of (2), and subtract it from the third, which will not alter the whole, and this gives

$$4ay = 4a^2x^2 - 4abx + b^2 + 4ac - b^2 = (2ax - b)^2 + ac - b^2. \quad (3)$$

Now it must be recollected that the square of any quantity is positive whether that quantity is positive or negative. This has been already sufficiently explained in saying that a change of the form of any expression does not change the form of its square. Common multiplication shows that $(c - d)^2$ and $(d - c)^2$ are the same thing; and, since one of these must be positive, the other must be also positive. Whenever, therefore, we wish to say that a quantity is positive, it can be done by supposing it equal to the square of an algebraical quantity. In equation (3) there are three distinct cases to be considered.

I. When b^2 is greater than $4ac$, that is, when $b^2 - 4ac$ is positive, let $b^2 - 4ac = k^2$, which expresses the condition.

Then

$$4ay = (2ax - b)^2 - k^2, \quad (4)$$

and we determine those values of x, which make $y = 0$, from the equation,

$$(2ax - b)^2 - k^2 = 0.$$

We have already solved such an equation, and we find that

$$2ax - b = \pm k,$$

where either sign may be taken. This shows that y or $ax^2 - bx + c$ is equal to nothing either when

$$\text{instead of } x \text{ is put} \quad \frac{b+k}{2a} = \frac{b+\sqrt{b^2-4ac}}{2a} = m,$$

$$\text{or} \quad \frac{b-k}{2a} = \frac{b-\sqrt{b^2-4ac}}{2a} = n,$$

the second values being formed by putting, instead of k its value $\sqrt{b^2 - 4ac}$. They are both positive quantities, because k^2 being equal to $b^2 - 4ac$ is less than b^2, and therefore k is less than b, and therefore $\frac{b+k}{2a}$ and $\frac{b-k}{2a}$ are both positive. These are the quantities which we have called m and n in the former investigations, and, therefore,

$$\begin{aligned} ax^2 - bx + c &= a(x-m)(x-n) \\ &= a\left(x - \frac{b+\sqrt{b^2-4ac}}{2a}\right)\left(x - \frac{b-\sqrt{b^2-4ac}}{2a}\right). \end{aligned}$$

Actual multiplication of the factors will show that this is an identical equation.

II. When b^2, instead of being greater than $4ac$, is equal to it, or when $b^2 - 4ac = 0$ and $k = 0$. In this case the values of m and n are equal, each being $\frac{b}{2a}$ and

$$y = ax^2 - bx + c = a(x-m)(x-n) = a\left(x - \frac{b}{2a}\right)^2.$$

In this case y is said, in algebra, to be a perfect square, since its square root can be extracted, and is $\sqrt{a}\left(x - \frac{b}{2a}\right)$. Arithmetically speaking, this would not be a perfect square unless a was a number whose square root could be extracted, but in algebra it is usual to call any quantity a perfect square with respect to any letter, which, when reduced, does not contain that letter under the sign $\sqrt{\ }$. This result is one which often occurs, and it must be recollected that when $b^2 - 4ac = 0$, $ax^2 - bx + c$ is a perfect square.

III. When b^2 is less than $4ac$, or when $b^2 - 4ac$ is negative and $4ac - b^2$ positive, let $4ac - b^2 = k^2$, and equation (3) becomes

$$4ay = (2ax - b)^2 + k^2.$$

In this case no value of x can ever make $y = 0$, for the equation $v^2 + w^2 = 0$ indicates that v^2 is equal to w^2 with a contrary sign, which cannot be, since all squares have the same sign. The values of x are said, in this case, to be impossible, and it indicates that there is something absurd or contradictory in the conditions of a problem which leads to such a result.

Having found that whenever

$$ax^2 - bx + c = a(x - m)(x - n),$$

it follows that $ax^2 + bx + c = a(x + m)(x + n)$, we know that

(1) when b^2 is greater than $4ac$,

$$ax^2+bx+c=a\left(x+\frac{b+\sqrt{b^2-4ac}}{2a}\right)\left(x+\frac{b-\sqrt{b^2-4ac}}{2a}\right);$$

(2) when $b^2=4ac$,

$$ax^2+bx+c=a\left(x+\frac{b}{2a}\right)^2,$$

and y is a perfect square;

(3) when b^2 is less than $4ac$, ax^2+bx+c cannot be divided into factors.

Now, let

$$y=ax^2-bx-c. \tag{1}$$

As before,

$$\begin{aligned} 4ay &= 4a^2x^2-4abx+b^2-4ac-b^2 \\ &= (2ax-b)^2-(b^2+4ac). \end{aligned} \tag{2}$$

Let $b^2+4ac=k^2$. Then

$$4ay=(2ax-b)^2-k^2. \tag{3}$$

Therefore y is 0 when $(2ax-b)^2=k^2$, or when $2ax-b=\pm k$.

That is,

$$m=\frac{b+k}{2a}=\frac{b+\sqrt{b^2+4ac}}{2a},$$

$$n=\frac{b-k}{2a}=\frac{b-\sqrt{b^2+4ac}}{2a}.$$

Now, because b^2 is less than $b^2 + 4ac$, b is less than $\sqrt{b^2 + 4ac}$, therefore n is a negative quantity. Leaving, for the present, the consideration of the negative quantity, we may decompose (3) into factors by means of the general formula

$$p^2 - q^2 = (p - q)(p + q),$$

which gives

$$\begin{aligned} 4ay &= (2ax - b - k)(2ax - b + k) \\ &= 4a^2\left(x - \frac{k + b}{2a}\right)\left(x + \frac{k - b}{2a}\right), \end{aligned}$$

from which y or

$$ax^2 - bx - c = a\left(x - \frac{\sqrt{b^2 + 4ac} + b}{2a}\right)\left(x + \frac{\sqrt{b^2 + 4ac} - b}{2a}\right).$$

Therefore, from what has been proved before,

$$ax^2 + bx - c = a\left(x + \frac{\sqrt{b^2 + 4ac} + b}{2a}\right)\left(x - \frac{\sqrt{b^2 + 4ac} - b}{2a}\right).$$

The following are some examples, of the truth of which the student should satisfy himself, both by reference to the ones just established, and by actual multiplication:

$$\begin{aligned} 2x^2 - 7x + 3 &= 2\left(x - \frac{7 + \sqrt{49 - 24}}{4}\right)\left(x - \frac{7 - \sqrt{49 - 24}}{4}\right) \\ &= 2(x - 3)(x - \tfrac{1}{2}), \end{aligned}$$

$$3x^2 - 6x + 1 = 3\left(x - \frac{3+\sqrt{6}}{3}\right)\left(x - \frac{3-\sqrt{6}}{3}\right),^*$$

$$5\tfrac{1}{2}x^2 - 22x + 22 = 5\tfrac{1}{2}(x-2)^2,$$

$$5x^2 + 9x - 7 = 5\left(x + \frac{\sqrt{221}+9}{10}\right)\left(x - \frac{\sqrt{221}-9}{10}\right).$$

If we collect together the different results at which we have arrived, to which species of tabulation the student should take care to accustom himself, we have the following:

$$ax^2 + bx + c$$
$$= a\left(x + \frac{b+\sqrt{b^2-4ac}}{2a}\right)\left(x + \frac{b-\sqrt{b^2-4ac}}{2a}\right), \quad (A)$$

$$ax^2 - bx + c$$
$$= a\left(x - \frac{b+\sqrt{b^2-4ac}}{2a}\right)\left(x - \frac{b-\sqrt{b^2-4ac}}{2a}\right), \quad (B)$$

$$ax^2 + bx - c$$
$$= a\left(x + \frac{\sqrt{b^2+4ac}+b}{2a}\right)\left(x - \frac{\sqrt{b^2+4ac}-b}{2a}\right), \quad (C)$$

*Recollect that $\sqrt{24} = \sqrt{6 \times 4} = \sqrt{6} \times \sqrt{4} = 2\sqrt{6}$.

$$ax^2 - bx - c$$
$$= a\left(x - \frac{\sqrt{b^2 + 4ac} + b}{2a}\right)\left(x + \frac{\sqrt{b^2 + 4ac} - b}{2a}\right). \quad (D)$$

These four cases may be contained in one, if we apply those rules for the change of signs which we have already established. For example, the first side of (C) is made from that of (A) by changing the sign of c; the second side of (C) is made from that of (A) in the same way. We have also seen the necessity of taking into account the negative quantities which satisfy an equation, as well as the positive ones; if we take these into account, each of the four forms of $ax^2 + bx + c$ can be made equal to nothing by two values of x. For example, in (1), when

$$ax^2 + bx + c = 0,$$
$$\text{either} \quad x + \frac{b - \sqrt{b^2 - 4ac}}{2a} = 0$$
$$\text{or} \quad x + \frac{b + \sqrt{b^2 - 4ac}}{2a} = 0.$$

If we call the values of x derived from the equations m and n, we find that

$$m = \frac{-b + \sqrt{b^2 - 4ac}}{2a}, \qquad n = \frac{-b - \sqrt{b^2 - 4ac}}{2a}. \quad (A')$$

In the cases marked (B), (C), and (D), the results are

$$m = \frac{b + \sqrt{b^2 - 4ac}}{2a}, \qquad n = \frac{b - \sqrt{b^2 - 4ac}}{2a}, \qquad (B')$$

$$m = \frac{-b + \sqrt{b^2 + 4ac}}{2a}, \qquad n = \frac{-b - \sqrt{b^2 + 4ac}}{2a}, \qquad (C')$$

$$m = \frac{b + \sqrt{b^2 + 4ac}}{2a}, \qquad n = \frac{b - \sqrt{b^2 + 4ac}}{2a}, \qquad (D')$$

and in all the four cases the form of $ax^2 + bx + c$ which is used, is the same as the corresponding form of

$$a(x - m)(x - n)$$

and the following results may be easily obtained. In (A') both m and n, if they exist at all, are negative. I say, if they exist at all, because it has been shown that if $b^2 - 4ac$ is negative, the quantity $ax^2 + bx + c$ cannot be divided into factors at all, since $\sqrt{b^2 - 4ac}$ is then no algebraical quantity, either positive or negative.

In (B') both, if they exist at all, are positive.

In (C') there are always real values for m and n, since $b^2 + 4ac$ is always positive; one of these values is positive, and the other negative, and the negative one is numerically the greatest.

In (D') there are also real values of m and n, one positive and the other negative, of which the positive one is numerically the greatest. Before proceeding any further, we must notice an extension of a phrase which is usually adopted.

The words *greater* and *less*, as applied to numbers, offer no difficulty, and from them we deduce, that if a be greater than b, $a - c$ is greater than $b - c$, as long as these subtractions are possible, that is, as long as c can be taken both from a and b. This is the only case which was considered when the rule was made, but in extending the meaning of the word *subtraction*, and using the symbol -3 to stand for $5-8$, the principle that if a be greater than b, $a-c$ is greater than $b - c$, leads to the following results. Since 6 is greater than 4, $6-12$ is greater than $4-12$, or -6 is greater than -8; again $6 - 6$ is greater than $4 - 6$, or 0 is greater than -2. These results, particularly the last, are absurd, as has been noticed, if we continue to mean by the terms *greater* and *less*, nothing more than is usually meant by them in arithmetic; but in extending the meaning of one term, we must extend the meaning of all which are connected with it, and we are obliged to apply the terms *greater* and *less* in the following way. Of two algebraical quantities with the same or different signs, that one is the greater which, when both are connected with a number numerically greater than either of them, gives the greater result. Thus -6 is said to be greater than -8, because $20 - 6$ is greater than $20 - 8$, 0 is greater than -4, because $6 + 0$ is greater than $6 - 4$; $+12$ is greater than -30, because $40 + 12$ is greater than $40 - 30$. Nevertheless -30 is said to be *numerically* greater than $+12$, because the number contained in the first is greater than that in the second. For this reason it was said, that in (C), the negative quantity was *numerically* greater, than the positive, because any positive quantity is in algebra called greater than any negative one, even though the number contained in the first should be less than that in the second. In the same way -14 is said to lie between $+3$ and -20, being less than the first and greater than the second. The advantage of these extensions is the same as that of others; the disadvantage attached to them, which it is not fair to disguise, is that, if used without proper caution, they lead the student into erroneous notions, which some elementary works, far from destroying, confirm, and even render necessary, by adopting these very notions as definitions; as for example, when they say that a negative quantity is one which is less than nothing; as if there could be such a thing, the usual meaning of the word less being considered, and as if the student had an idea of a quantity less than nothing already in his mind, to which it was only necessary to give a name.

The product $(x-m)(x-n)$ is positive when $(x-m)$ and $(x-n)$ have the same, and negative when they have different signs. This last can never happen except when x lies between m and n, that is, when x is algebraically greater than the one, and less than the other. The following table will exhibit this, where different products are taken with various signs of m and n, and three values are given to x one after the other, the first of which is less than both m and n, the second between both, and the third greater than both.

The student will see the reason of this, and perform a useful exercise in making two or three tables of this description for himself. The result is that $(x-m)(x-n)$ is negative

PRODUCT.	VALUE OF x.	VALUE OF THE PRODUCT WITH ITS SIGN.
$(x-4)(x-7)$	$+1$	$+18$
$m=+4$	$+5$	-2
$n=+7$	$+10$	$+18$
$(x+10)(x-3)$	-12	$+30$
$m=-10$	-7	-30
$n=+3$	$+4$	$+14$
$(x+2)(x+12)$	-13	$+11$
$m=-2$	-6	-24
$n=-12$	-1	$+11$

when x lies between m and n; is nothing when x is either equal to m or to n, and positive when x is greater than both, or less than both. Consequently, $a(x-m)(x-n)$ has the same sign as a when x is greater than both m and n, or less than both, and a different sign from a when x lies between both. But whatever may be the signs of a, b, and c, if there are two quantities m and n, which make

$$ax^2+bx+c=a(x-m)(x-n),$$

that is, if the equation $ax^2+bx+c=0$ has real roots, the expression ax^2+bx+c always has the same sign as a for all values of x, except when x lies between these roots.

It only remains to consider those cases in which ax^2+bx+c cannot be decomposed into different factors, which happens whenever $b^2 - 4ac$ is 0, or negative. In the first case when $b^2 - 4ac = 0$, we have

$$ax^2 + bx + c = a\left(x + \frac{b}{2a}\right)^2,$$
$$ax^2 - bx + c = a\left(x - \frac{b}{2a}\right)^2,$$

and as these expressions are composed of factors, one of which is a square, and therefore positive, they have always the same sign as the other factor, which is a. When $b^2 - 4ac$ is negative, we have proved that if $y = ax^2 \pm bx + c$, $4ay = (2ax \pm b)^2 + k^2$, where $k^2 = 4ac - b^2$, and therefore $4ay$ being the sum of two squares is always positive, that is, $ax^2 \pm bx + c$ has the same sign as a, whatever may be the value of x. When $c = 0$, the expression becomes $ax^2 + bx$, or $x(ax + b)$, which is nothing either when $x = 0$, or when $ax + b = 0$ and $x = -\frac{b}{a}$; the general expressions for m and n become in this case $\frac{-b + \sqrt{b^2}}{2a}$ and $\frac{-b - \sqrt{b^2}}{2a}$, which give the same results.

When $b = 0$, the expression is reduced to $ax^2 + c = 0$, which is nothing when $x = \pm\sqrt{-\frac{c}{a}}$, which is not possible, except when c and a have different signs. In this case, that is, when the expression assumes the form $ax^2 - c$, it is the same as

$$a\left(x-\sqrt{\frac{c}{a}}\right)\left(x+\sqrt{\frac{c}{a}}\right).$$

The same result might be deduced by making $b=0$ in the general expressions for m and n.

When $a=0$, the expression is reduced to $bx+c$, which is made equal to nothing by one value of x only, that is $-\frac{c}{b}$. If we take the general expressions for m and n, and make $a=0$ in them, that is, in $\frac{-b+\sqrt{b^2-4ac}}{2a}$, and $\frac{-b-\sqrt{b^2-4ac}}{2a}$, we find as the results $\frac{0}{0}$ and $-\frac{2b}{0}$. These have been already explained. The first may either indicate that any value of x will solve the problem which produced the equation $ax^2+bx+c=0$, or that we have applied a rule to a case which was not contemplated in its formation, and have thereby created a factor in the numerator and denominator of x, which, in attempting to apply the rule, becomes equal to nothing. The student is referred to the problem of the two couriers, solved in the preceding part of this treatise. The latter is evidently the case here, because in returning to the original equation, we find it reduced to $bx+c=0$, which gives a rational value for x, namely, $-\frac{c}{b}$. The second value, or $-\frac{2b}{0}$, which in algebraical language is called infinite, may indicate, that though there is no other value of x, except $-\frac{c}{b}$, which solves the equation, still that the greater the number which is taken for x, the more nearly is a second solution obtained. The use of these expressions is to point out the cases in which there is anything remarkable in the general problem; to the problem itself we must resort for further explanation.

The importance of the investigations connected with the expression $ax^2 + bx + c$, can hardly be overrated, at least to those students who pursue mathematics to any extent. In the higher branches, great familiarity with these results is indispensable. The student is therefore recommended not to proceed until he has completely mastered the details here given, which have been hitherto too much neglected in English works on algebra.

In solving equations of the second degree, we have obtained a new species of result, which indicates that the problem cannot be solved at all. We refer to those results which contain the square root of a negative quantity. We find that by multiplication the squares of $c - d$ and of $d - c$ are the same, both being $c^2 - 2cd + d^2$. Now either $c - d$ or $d - c$ is positive, and since they both have the same square, it appears that the squares of all quantities, whether positive or negative, are positive. It is therefore absurd to suppose that there is any quantity which x can represent, and which satisfies the equation $x^2 = -a^2$, since that would be supposing that x^2, a positive quantity, is equal to the negative quantity $-a^2$. The solution is then said to be impossible, and it will be easy to show an instance in which such a result is obtained, and also to show that it arises from the absurdity of the problem.

Let a number a be divided into any two parts, one of which is greater than the half, and the other less. Call the first of these $\frac{a}{2}+x$, then the second must be $\frac{a}{2}-x$, since the sum of both parts must be a. Multiply these parts together, which gives

$$\left(\frac{a}{2}+x\right)\left(\frac{a}{2}-x\right), \quad \text{or} \quad \left(\frac{a}{2}\right)^2 - x^2.$$

As x diminishes, this product increases, and is greatest of all when $x = 0$, that is, when the two parts, into which a is divided, are $\frac{a}{2}$ and $\frac{a}{2}$, or when the number a is halved. In this case the product of the parts is $\frac{a}{2} \times \frac{a}{2}$, or $\frac{a^2}{4}$, and a number a can never be divided into two parts whose product is greater than $\frac{a^2}{4}$. This being premised, suppose that we attempt to divide the number a into two parts, whose product is b. Let x be one of these parts, then $a - x$ is the other, and their product is $ax - x^2$.

We have, therefore,

$$ax - x^2 = b,$$
$$\text{or} \quad x^2 - ax + b = 0.$$

If we solve this equation, the two roots are the two parts required, since from what we have proved of the expression

$x^2 - ax + b$ the sum of the roots is a and their product b. These roots are

$$\frac{a}{2} + \sqrt{\frac{a^2}{4} - b} \quad \text{and} \quad \frac{a}{2} - \sqrt{\frac{a^2}{4} - b},^*$$

which are impossible when $\dfrac{a^2}{4} - b$ is negative, or when b is greater than $\dfrac{a^2}{4}$, which agrees with what has just been proved, that no number is capable of being divided into two parts whose product is greater than $\dfrac{a^2}{4}$.

We have shown the symbol $\sqrt{-a}$ to be void of meaning, or rather self-contradictory and absurd. Nevertheless, by means of such symbols, a part of algebra is established which is of great utility. It depends upon the fact, which must be verified by experience, that the common rules of algebra may be applied to these expressions without leading to any false results. An appeal to experience of this nature appears to be contrary to the first principles laid down at the beginning of this work. We cannot deny that it is so in reality, but it must be recollected that this is but a small and isolated part of an immense subject, to all other branches of which these principles apply in their fullest extent. There have not been wanting some to assert that these symbols may be used as

---

*The general expressions for m and n give $\dfrac{a \pm \sqrt{a^2 - 4b}}{2}$ as the roots of $x^2 - ax + b = 0$.

rationally as any others, and that the results derived from them are as conclusive as any reasoning could make them. I leave the student to discuss this question as soon as he has acquired sufficient knowledge to understand the various arguments: at present let him proceed with the subject as a part of the mechanism of algebra, on the assurance that by careful attention to the rules laid down he can never be led to any incorrect result. The simple rule is, apply all those rules to such expressions as $\sqrt{-a}$, $a+\sqrt{-b}$, etc., which have been proved to hold good for such quantities as $\sqrt{a}$, $a+\sqrt{b}$, etc. Such expressions as the first of these are called *imaginary*, to distinguish them from the second, which are called *real*; and it must always be recollected that there is no quantity, either positive or negative, which an imaginary expression can represent.

It is usual to write such symbols as $\sqrt{-b}$ in a different form. To the equation $-b = b \times (-1)$ apply the rule derived from the equation $\sqrt{xy} = \sqrt{x} \times \sqrt{y}$, which gives $\sqrt{-b} = \sqrt{b} \times \sqrt{-1}$, of which the first factor is real and the second imaginary. Let $\sqrt{b} = c$, then $\sqrt{-b} = c\sqrt{-1}$. In this way all expressions may be so arranged that $\sqrt{-1}$ shall be the only imaginary quantity which appears in them. Of this reduction the following are examples:

$$\sqrt{-24} = \sqrt{24}\sqrt{-1} = 2\sqrt{6}\sqrt{-1},$$

$$\sqrt{-a^2} = a\sqrt{-1},$$

$$\sqrt{-a} \times \sqrt{-a} = -a,$$

$$\sqrt{2ab - a^2 - b^2} = (a - b)\sqrt{-1},$$
$$\sqrt{-a^2} \times \sqrt{-b^2} = a\sqrt{-1} \times b\sqrt{-1} = -ab.$$

The following tables exhibit other applications of the rules:

$$\begin{aligned} c &= a\sqrt{-1} & c^7 &= -a^7\sqrt{-1}, \\ c^2 &= -a^2 & c^8 &= a^8, \quad \text{etc.}, \\ c^3 &= -a^3\sqrt{-1} & c^{4n-3} &= a^{4n-3}\sqrt{-1}, \\ c^4 &= a^4 & c^{4n-2} &= -a^{4n-2}, \\ c^5 &= a^5\sqrt{-1} & c^{4n-1} &= -a^{4n-1}\sqrt{-1}, \\ c^6 &= -a^6 & c^{4n} &= a^{4n}. \end{aligned}$$

The powers of such an expression as $a\sqrt{-1}$ are therefore alternately real and imaginary, and are positive and negative in pairs.

$$(a + b\sqrt{-1})^2 = a^2 - b^2 + 2ab\sqrt{-1},$$
$$(a - b\sqrt{-1})^2 = a^2 - b^2 - 2ab\sqrt{-1},$$
$$(a + b\sqrt{-1})(a - b\sqrt{-1}) = a^2 + b^2,$$
$$\frac{a + b\sqrt{-1}}{a - b\sqrt{-1}} = \frac{a^2 - b^2}{a^2 + b^2} + \frac{2ab\sqrt{-1}}{a^2 + b^2},$$
$$(a + b\sqrt{-1})(c + d\sqrt{-1}) = ac - bd + (ad + bc)\sqrt{-1}.$$

Let the roots of the equation $ax^2 + bx + c$ be impossible, that is, let $b^2 - 4ac$ be negative and equal to $-k^2$. Its roots, as derived from the rules established when b^2-4ac was positive, are

$$\frac{-b+\sqrt{-k^2}}{2a} \quad \text{and} \quad \frac{-b-\sqrt{-k^2}}{2a}, \quad \text{or}$$

$$-\frac{b}{2a}+\frac{k}{2a}\sqrt{-1} \quad \text{and} \quad -\frac{b}{2a}-\frac{k}{2a}\sqrt{-1}.$$

Take either of these instead of x; for example, let

$$x=-\frac{b}{2a}+\frac{k}{2a}\sqrt{-1}.$$

$$\text{Then}\quad ax^2=\frac{b^2}{4a}-\frac{bk}{2a}\sqrt{-1}-\frac{k^2}{4a},$$

$$bx=-\frac{b^2}{2a}+\frac{bk}{2a}\sqrt{-1},$$

$$c=c.$$

Therefore, $ax^2+bx+c=\frac{b^2}{4a}-\frac{k^2}{4a}-\frac{b^2}{4a}+c$, in which, if $4ac-b^2$ be substituted instead of k^2, the result is 0. It appears, then, that the imaginary expressions which take the place of the roots when b^2-4ac is negative, will, if the ordinary rules be applied, produce the same results as the roots. They are thence called imaginary roots, and we say that every equation of the second degree has two roots, either both real or both imaginary. It is generally true, that wherever an imaginary expression occurs the same results will follow from the application of these expressions in any process as would have followed had the proposed problem been possible and its solution real.

When an equation arises in which imaginary and real expressions occur together, such as $a + b\sqrt{-1} = c + d\sqrt{-1}$, when all the terms are transferred on one side, the part which is real and that which is imaginary must each of them be equal to nothing. The equation just given when its left side is transposed becomes $a - c + (b - d)\sqrt{-1} = 0$. Now, if b is not equal to d, let $b - d = e$; then $a - c + e\sqrt{-1} = 0$, and $\sqrt{-1} = \dfrac{c - a}{e}$; that is, an imaginary expression is equal to a real one, which is absurd. Therefore, $b = d$ and the original equation is thereby reduced to $a = c$. This goes on the supposition that a, b, c, and d are real. If they are not so there is no necessary absurdity in $\sqrt{-1} = \dfrac{c - a}{e}$. If, then, we wish to express that two possible quantities a and b are respectively equal to two others c and d, it may be done at once by the equation

$$a + b\sqrt{-1} = c + d\sqrt{-1}.$$

The imaginary expression $\sqrt{-a}$ and the negative expression $-b$ have this resemblance, that either of them occurring as the solution of a problem indicates some inconsistency or absurdity. As far as real meaning is concerned, both are equally imaginary, since $0 - a$ is as inconceivable as $\sqrt{-a}$. What, then, is the difference of signification? The following problems will elucidate this. A father is fifty-six, and his son twenty-nine years old: when will the father be twice as old as the son? Let this happen x years from the present time; then the age of the father will be $56+x$, and that of the son $29+x$; and therefore, $56+x = 2(29+x) = 58+2x$, or $x = -2$. The result is absurd; nevertheless, if in the equation we change the sign of x throughout it becomes $56 - x = 58 - 2x$, or $x = 2$. This equation is the one belonging to the problem: a father is 56 and his son 29 years old; when *was* the father twice as old as the son? the answer to which is, two years ago. In this case the negative sign arises from too great a limitation in the terms of the problem, which should have demanded how many years have elapsed or will elapse before the father is twice as old as his son?

Again, suppose the problem had been given in this last-mentioned way. In order to form an equation, it will be necessary either to suppose the event past or future. If of the two suppositions we choose the wrong one, this error will be pointed out by the negative form of the result. In this case the negative result will arise from a mistake in reducing the problem to an equation. In either case, however, the result may be interpreted, and a rational answer to the question may be given. This, however, is not the case in a problem, the result of which is imaginary. Take the instance above solved, in which it is required to divide a into two parts, whose product is b. The resulting equation is

$$x^2 - ax + b = 0, \quad \text{or} \quad x = \frac{a}{2} \pm \sqrt{\frac{a^2}{4} - b},$$

the roots of which are imaginary when b is greater than $\frac{a^2}{4}$. If we change the sign of x in the equation it becomes

$$x^2 + ax + b = 0, \quad \text{or} \quad x = -\frac{a}{2} \pm \sqrt{\frac{a^2}{4} - b},$$

and the roots of the second are imaginary, if those of the first are so. There is, then, this distinct difference between the negative and the imaginary result. When the answer to a problem is negative, by changing the sign of x in the equation which produced that result, we may either discover an error in the method of forming that equation or show that the question of the problem is too limited, and may be extended so as to admit of a satisfactory answer. When the answer to a problem is imaginary this is not the case.

CHAPTER XI.

ON ROOTS IN GENERAL, AND LOGARITHMS.

THE meaning of the terms *square root*, *cube root*, *fourth root*, etc., has already been defined. We now proceed to the difficulties attending the connexion of the roots of a with the powers of a. The following table will refresh the memory of the student with respect to the meaning of the terms:

NAME OF x.		NAME OF x.	
Square of a	$x = aa$	Square Root of a	$xx = a$
Cube of a	$x = aaa$	Cube Root of a	$xxx = a$
Fourth Power of a	$x = aaaa$	Fourth Root of a	$xxxx = a$
Fifth Power of a	$x = aaaaa$	Fifth Root of a	$xxxxx = a$

The different powers and roots of a have hitherto been expressed in the following way:

$$\text{Powers}\quad a^2 \quad a^3 \quad a^4 \quad a^5 \;\ldots\; a^m \;\ldots\; a^{m+n}, \quad \text{etc.}$$

$$\text{Roots}\quad \sqrt[2]{a}^{*} \quad \sqrt[3]{a} \quad \sqrt[4]{a} \quad \sqrt[5]{a} \quad \sqrt[m]{a} \quad \sqrt[m+n]{a}, \quad \text{etc.}$$

which series are connected together by the following equation, $(\sqrt[n]{a})^n = a$.

There has hitherto been no connexion between the manner of expressing powers and roots, and we have found no properties which are common both to powers and roots. Nevertheless, by the extension of rules, we shall be led to

*The 2 is usually omitted, and the square root is written thus $\sqrt{a}$.

a method of denoting the raising of powers, the extraction of roots, and combinations of the two, to which algebra has been most peculiarly indebted, and the importance of which will justify the length at which it will be treated here.

Suppose it required to find the cube of $2a^2b^3$; that is, to find $2a^2b^3 \times 2a^2b^3 \times 2a^2b^3$. The common rules of multiplication give, as the result, $8a^6b^9$, which is expressed in the following equation,

$$(2a^2b^3)^3 = 8a^6b^9.$$

Similarly

$$(3a^4b^3)^4 = 81a^{16}b^{12},$$
$$\left(\frac{1}{2}\frac{b^4}{a}\right)^6 = \frac{1}{64}\frac{b^{24}}{a^6};$$

and the general rule by which any single term may be raised to the power whose index is n, is: Raise the coefficient to the power n, and multiply the index of every letter by n, that is,

$$(a^pb^qc^r)^n = a^{np}b^{nq}c^{nr}.$$

In extracting the root of any simple term, we are guided by the manner in which the corresponding power is found. The rule is: Extract the required root of the coefficient, and divide the index of each letter by the index of the root. Where these divisions do not give whole numbers as the quotients, the expression whose root is to be extracted does not admit of the extraction without the introduction of some new symbol. For example, extract the fourth root of $16a^{12}b^8c^4$, or find $\sqrt[4]{16a^{12}b^8c^4}$. The expression here given is the same as the following:

$$2a^3b^2c \times 2a^3b^2c \times 2a^3b^2c \times 2a^3b^2c,$$

or $(2a^3b^2c)^4$, the fourth root of which is $2a^3b^2c$, conformably to the rule.

Any root of a product, such as AB, may be extracted by extracting the root of each of its factors. Thus, $\sqrt[3]{AB} = \sqrt[3]{A}\sqrt[3]{B}$. For, raise $\sqrt[3]{A}\sqrt[3]{B}$ to the third power, the result of which is,

$$\sqrt[3]{A}\sqrt[3]{B} \times \sqrt[3]{A}\sqrt[3]{B} \times \sqrt[3]{A}\sqrt[3]{B},$$

$$\text{or} \quad \sqrt[3]{A}\sqrt[3]{A}\sqrt[3]{A} \times \sqrt[3]{B}\sqrt[3]{B}\sqrt[3]{B},$$

$$\text{or} \quad AB.$$

In the same way it may be proved generally, that $\sqrt[n]{ABC} = \sqrt[n]{A}\sqrt[n]{B}\sqrt[n]{C}$. The most simple way of representing any root of any expression is the dividing it into two factors, one of which is the highest which it admits of whose root can be extracted by the rule just given. For example, in finding $\sqrt[3]{16a^4b^7c}$ we must observe that 16 is 8×2, a^4 is $a^3 \times a$, b^7 is $b^6 \times b$, and the expression is $8a^3b^6 \times 2abc$, the cube root of which, found by extracting the cube root of each factor, is $2ab^2\sqrt[3]{2abc}$. The second factor has no cube root which can be expressed by means of the symbols hitherto used, but when the numbers which a, b, and c stand for are known, $\sqrt[3]{2abc}$ maybe found either exactly, or, when that is not possible, by approximation.

We find that a power of a power is found by affixing, as an index, the product of the indices of the two powers. Thus $(a^2)^4$ or $a^2 \times a^2 \times a^2 \times a^2$ is a^8, or $a^{4\times 2}$. This is the same as $(a^4)^2$, which is $a^4 \times a^4$, or a^8. Therefore, generally $(a^m)^n = (a^n)^m = a^{mn}$. In the same manner, a root of a root is the root whose index is the product of the indices of the two roots. Thus

$$\sqrt[3]{\sqrt[2]{a}} = \sqrt[6]{a}.$$

For since $a = \sqrt[6]{a}\sqrt[6]{a}\sqrt[6]{a} \times \sqrt[6]{a}\sqrt[6]{a}\sqrt[6]{a}$, the square root of a is $\sqrt[6]{a}\sqrt[6]{a}\sqrt[6]{a}$, the cube root of which is $\sqrt[3]{a}$. This is the same as $\sqrt[2]{\sqrt[3]{a}}$, and generally

$$\sqrt[m]{\sqrt[n]{a}} = \sqrt[n]{\sqrt[m]{a}} = \sqrt[mn]{a}.$$

Again, when a power is raised and a root extracted, it is indifferent which is done first. Thus $\sqrt[3]{a^2}$ is the same thing as $(\sqrt[3]{a})^2$. For since $a^2 = a \times a$, the cube root may be found by taking the cube root of each of these factors, that is $\sqrt[3]{a^2} = \sqrt[3]{a} \times \sqrt[3]{a} = (\sqrt[3]{a})^2$, and generally

$$\sqrt[n]{a^m} = (\sqrt[n]{a})^m.$$

In the expression $\sqrt[n]{a^m}$, n and m may both be multiplied by any number, without altering the expression, that is

$$\sqrt[np]{a^{mp}} = (\sqrt[n]{a})^m.$$

To prove this, recollect that

$$\sqrt[np]{a^{mp}} = \sqrt[n]{\sqrt[p]{a^{mp}}}.$$

But a^{mp} is $(a^m)^p$, and by definition $\sqrt[p]{(a^m)^p} = a^m$. Therefore $\sqrt[np]{a^{mp}} = \sqrt[n]{a^m}$. This multiplication is equivalent to raising a power of $\sqrt[n]{a^m}$, and afterwards reducing the result to its former value, by extracting the corresponding root, in the same way as $\frac{mp}{np}$ signifies that $\frac{m}{n}$ has been multiplied by p, and the result has been restored to its former value by dividing it by p.

The following equations should be established by the student to familiarise him with the notation and principles hitherto laid down.

$$\sqrt[n]{(a-b)^{n-2}} \times \sqrt[3n]{(a-b)^6} = a - b,$$

$$\sqrt[n+m]{(a+b)^{n-m}} \times \sqrt[n-m]{(a-b)^{n+m}} = (a^2 - b^2)\left(\frac{\sqrt[n-m]{a-b}}{\sqrt[n+m]{a+b}}\right)^{2m},$$

$$\sqrt[n]{\frac{ab}{cd}} = \frac{\sqrt[n]{ab}}{\sqrt[n]{cd}} = \frac{\sqrt[n]{a}\,\sqrt[n]{b}}{\sqrt[n]{c}\,\sqrt[n]{d}} = \sqrt[n]{\frac{a}{c}} \times \sqrt[n]{\frac{b}{d}},$$

$$\sqrt[n]{\frac{a}{b}} = \frac{\sqrt[n]{ab^{n-1}}}{b} = \frac{a}{\sqrt[n]{a^{n-1}b}}.$$

The quantity $\sqrt[n]{a^m}$ is a simple expression when m can be divided by n, without remainder, for example $\sqrt[2]{a^{12}} = a^6$, $\sqrt[5]{a^{20}} = a^4$, and in general, whenever m can be divided by n

without remainder, $\sqrt[n]{a^m} = a^{\frac{m}{n}}$. This symbol, viz., a letter which has an exponent appearing in a fractional form, has not hitherto been used. We may give it any meaning which we please, provided it be such that when $\dfrac{m}{n}$ is fractional in form only, and not in reality, that is, when m is divisible by n, and the quotient is p, $a^{\frac{m}{n}}$ shall stand for a^p, or $aaa\ldots(p)$.* It will be convenient to let $a^{\frac{m}{n}}$ always stand for $\sqrt[n]{a^m}$, in which case the condition alluded to is fulfilled, since when $\dfrac{m}{n} = p$, $a^{\frac{m}{n}}$ or $\sqrt[n]{a^m} = a^p$. This extension of a rule, the advantages of which will soon be apparent, is exemplified in the following table, which will familiarise the student with the different cases of this new notation:

$$a^{\frac{1}{2}} \text{ stands for } \sqrt[2]{a^1} \text{ or } \sqrt{a},$$

$$a^{\frac{1}{3}} \text{ stands for } \sqrt[3]{a},$$

$$a^{\frac{1}{4}} \text{ stands for } \sqrt[4]{a},$$

$$a^{\frac{2}{3}} \text{ stands for } \sqrt[3]{a^2} \text{ or } (\sqrt[3]{a})^2,$$

$$a^{\frac{7}{5}} \text{ stands for } \sqrt[5]{a^7} \text{ or } (\sqrt[5]{a})^7,$$

$$a^{\frac{m+n}{m-n}} \text{ stands for } \sqrt[m-n]{a^{m+n}},$$

---

*This is a notation in common use, and means that $aaa\ldots$ is to be continued until it has been repeated p times. Thus

$$a + a + a + \ldots(p) = pa,$$
$$a \times a \times a \times \ldots(p) = a^p.$$

$$(p+q)^{\frac{m-n}{2}} \text{ stands for } \sqrt{(p+q)^{m-n}},$$
$$(c^{\frac{m}{n}})^{\frac{p}{q}} \text{ stands for } \sqrt[q]{(\sqrt[n]{c^m})^p},$$
$$(a^{\frac{1}{n}})^{\frac{1}{q}} \text{ stands for } \sqrt[q]{\sqrt[n]{a}}.$$

The results at which we have arrived in this chapter, translated into this new language, are as follows:

$$(x^{\frac{1}{n}})^n = (x^n)^{\frac{1}{n}} = x, \tag{1}$$
$$(ABC)^{\frac{1}{n}} = A^{\frac{1}{n}}B^{\frac{1}{n}}C^{\frac{1}{n}}, \tag{2}$$
$$(a^{\frac{1}{n}})^{\frac{1}{q}} = a^{\frac{1}{nq}}, \tag{3}$$
$$(a^m)^{\frac{1}{n}} = (a^{\frac{1}{n}})^m = a^{\frac{m}{n}}, \tag{4}$$
$$a^{\frac{m}{n}} = a^{\frac{mp}{nq}}. \tag{5}$$

The advantages resulting from the adoption of this notation, are, (1) that time is saved in writing algebraical expressions; (2) all rules which have been shown to hold good for performing operations upon such quantities as a^m, hold good also for performing the same operations upon such quantities as $a^{\frac{m}{n}}$, in which the exponents are fractional. The truth of this last assertion we proceed to establish.

Suppose it required to multiply together $a^{\frac{m}{n}}$ and $a^{\frac{l}{n}}$, or $\sqrt[n]{a^m}$ and $\sqrt[n]{a^l}$. From (2) this is $\sqrt[n]{a^m \times a^l}$, or $\sqrt[n]{a^{m+l}}$, or $a^{\frac{m+l}{n}}$. Suppose it now required to multiply $a^{\frac{m}{n}}$ and $a^{\frac{p}{q}}$. From (5) the first of these is the same as $a^{\frac{mq}{nq}}$, and the second is the same as $a^{\frac{np}{nq}}$. The product of these by the last case is $a^{\frac{mq+np}{nq}}$, or $\sqrt[nq]{a^{mq+np}}$. But $\dfrac{mq+np}{nq}$ is $\dfrac{m}{n}+\dfrac{p}{q}$, and therefore

$$a^{\frac{m}{n}} \times a^{\frac{p}{q}} = a^{\frac{m}{n}+\frac{p}{q}}. \tag{6}$$

This is the same result as was obtained when the indices were whole numbers. The rule is: To multiply together two powers of the same quantity, add the indices, and make the sum the index of the product. It follows in the same way that

$$\frac{a^{\frac{m}{n}}}{a^{\frac{p}{q}}} = a^{\frac{m}{n}-\frac{p}{q}} = a^{\frac{mq-np}{nq}} = \sqrt[nq]{a^{mq-np}},$$

or, to divide one power of a quantity by another, subtract the index of the divisor from that of the dividend, and make the difference the index of the result.

Suppose it required to find $(a^{\frac{m}{n}})^p$. It is evident that $a^{\frac{m}{n}} \times a^{\frac{m}{n}} = a^{\frac{m}{n}+\frac{m}{n}} = a^{\frac{2m}{n}}$, or $(a^{\frac{m}{n}})^2 = a^{\frac{2m}{n}}$. Similarly $(a^{\frac{m}{n}})^3 = a^{\frac{3m}{n}}$, and so on. Therefore $(a^{\frac{m}{n}})^p = a^{\frac{mp}{n}}$.

Again to find $(a^{\frac{m}{n}})^{\frac{1}{q}}$, or $\sqrt[q]{a^{\frac{m}{n}}}$. Let this be $a^{\frac{x}{y}}$. Then $a^{\frac{x}{y}} = \sqrt[q]{a^{\frac{m}{n}}}$, or $(a^{\frac{x}{y}})^q = a^{\frac{m}{n}}$, or $a^{\frac{xq}{y}} = a^{\frac{m}{n}}$. Therefore $\dfrac{xq}{y} = \dfrac{m}{n}$, or $\dfrac{x}{y} = \dfrac{m}{nq}$, and $(a^{\frac{m}{n}})^{\frac{1}{q}} = a^{\frac{m}{nq}}$.

Again to find $(a^{\frac{m}{n}})^{\frac{p}{q}}$, or $\sqrt[q]{(a^{\frac{m}{n}})^p}$. Apply the last two rules, and it appears that $(a^{\frac{m}{n}})^p = a^{\frac{mp}{n}}$, and $\sqrt[q]{a^{\frac{mp}{n}}} = a^{\frac{mp}{nq}}$. Therefore $(a^{\frac{m}{n}})^{\frac{p}{q}} = a^{\frac{mp}{nq}} = a^{\frac{m}{n}\times\frac{p}{q}}$.

The rule is: To raise one power of a quantity to another power, multiply the indices of the two powers together, and make the product the index of the result. All these rules are exactly those which have been shown to hold good when the indices are whole numbers. But there still remains one remarkable extension, which will complete this subject.

We have proved that whether m and n be whole or fractional numbers, $\frac{a^n}{a^m} = a^{m-n}$. The only cases which have been considered in forming this rule are those in which m is greater than n, being the only ones in which the subtraction indicated is possible. If we apply the rule to any other case, a new symbol is produced, which we proceed to consider. For example, suppose it required to find $\frac{a^3}{a^7}$. If we apply the rule, we find the result a^{3-7}, or a^{-4}, for which we have hitherto no meaning. As in former cases, we must apply other methods to the solution of this case, and when we have obtained a rational result, a^{-4} may be used in future to stand for this result. Now the fraction $\frac{a^3}{a^7}$ is the same as $\frac{1}{a^4}$, which is obtained by dividing both its numerator and denominator by a^3. Therefore $\frac{1}{a^4}$ is the rational result, for which we have obtained a^{-4} by applying a rule in too extensive a manner. Nevertheless, if a^{-4} be made to stand for $\frac{1}{a^4}$, and a^{-m} for $\frac{1}{a^m}$, the rule will always give correct results, and the general rules for multiplication, division, and raising of powers remain the same as before. For example, $a^{-m} \times a^{-n}$

is $\frac{1}{a^m} \times \frac{1}{a^n}$, or $\frac{1}{a^m a^n}$, which is $\frac{1}{a^{m+n}}$, or $a^{-(m+n)}$, or a^{-m-n}. Similarly

$$\frac{a^{-m}}{a^{-n}}, \text{ or } \frac{\frac{1}{a^m}}{\frac{1}{a^n}}, \text{ is } \frac{a^n}{a^m}, \text{ or } a^{n-m}, \text{ or } a^{-m-(-n)}.$$

Again

$$(a^m)^{-n} \text{ is } \frac{1}{(a^m)^n}, \text{ or } \frac{1}{a^{mn}}, \text{ or } a^{-mn},$$

and so on.

It has before been shown that a^0 stands for 1 whenever it occurs in the solution of a problem. We can now, therefore, assign a meaning to the expression a^m, whether m be whole or fractional, positive, negative, or nothing, and in all these cases the following rules hold good:

$$a^m \times a^n = a^{m+n},$$
$$\frac{a^m}{a^n} = a^{m-n} = a^m a^{-n},$$
$$(a^m)^n = (a^n)^m = a^{mn}.$$

The student can now understand the meaning of such an expression as $10^{\cdot 301}$, where the index or exponent is a decimal fraction. Since .301 is $\frac{301}{1000}$, this stands for $\sqrt[1000]{(10)^{301}}$, an expression of which it would be impossible to calculate the value by any method which the student has hitherto been taught, but which may be shown by other processes to be very nearly equal to 2.

Before proceeding to the practice of logarithmic calculations, the student should thoroughly understand the meaning of fractional and negative indices, and be familiar with the operations performed by means of them. He should work many examples of multiplication and division in which they occur, for which he can have recourse to any elementary work. The rules are the same as those to which he has been accustomed, substituting the addition, subtraction, and so forth, of fractional indices, instead of these which are whole numbers.

In order to make use of logarithms, he must provide himself with a table. Either of the following works may be recommended to him:

[1. Bruhns, *A New Manual of Logarithms to Seven Places of Decimals* (English preface, Leipsic).

2. Schrön, *Seven-Figure Logarithms* (English edition, London).

3. Bremiker's various editions of *Vega's Logarithmic Tables* (Weidmann, Berlin). With English preface.

4. Callet, *Tables portatives de Logarithmes.* (Last impression, Paris, 1890).

5. V. Caillet, *Tables des Logarithmes et Co-Logarithmes des nombres* (Paris).

6. Hutton's *Mathematical Tables* (London).

7. Chambers's *Mathematical Tables* (Edinburgh).

8. The American six-figure Tables of Jones, of Wells, and

of Haskell.

For fuller bibliographical information on the subject of tables of logarithms, see the *Encyclopædia Britannica*, Article "Tables," Vol. XXIII.—*Ed.*]*

The limits of this treatise will not allow us to enter into the subject of the definition, theory, and use of logarithms, which will be found fully treated in the standard text-books of Arithmetic, Algebra, and Trigonometry. There is, however, one consideration connected with the tables, which, as it involves a principle of frequent application, it will be well to explain here. On looking into any table of logarithms it will be seen, that for a series of numbers the logarithms increase in arithmetical progression, as far as the first seven places of decimals are concerned; that is, the difference between the successive logarithms continues the same. For ex-

---

*The original text of De Morgan, for which the above paragraph has been substituted, reads as follows: "Either of the following works may be recommended to him: (1) Taylor's *Logarithms.* (2) Hutton's *Logarithms.* (3) Babbage, *Logarithms of Numbers*; Callet, *Logarithms of Sines, Cosines*, etc. (4) Bagay, *Tables Astronomiques et Hydrographiques.* The first and last of these are large works, calculated for the most accurate operations of spherical trigonometry and astronomy. The second and third are better suited to the ordinary student. For those who require a pocket volume there are Lalande's and Hassler's Tables, the first published in France, the second in the United States."—*Ed.*

ample, the following is found from any tables:

$$\log 41713 = 4.6202714,$$
$$\log 41714 = 4.6202818,$$
$$\log 41715 = 4.6202922.$$

The difference of these successive logarithms and of almost all others in the same page is .0000104. Therefore in this the addition of 1 to the number gives an addition of .0000104 to the logarithm. It is a general rule that when one quantity depends for its value upon another, as a logarithm does upon its number, or an algebraical expression, such as x^2+x upon the letter or letters which it contains, if a very small addition be made to the value of one of these letters, in consequence of which the expression itself is increased or diminished; generally speaking, the increment* of the expression will be very nearly proportional to the increment of the letter whose value is increased, and the more nearly so the smaller is the increment of the letter. We proceed to illustrate this. The product of two fractions, each of which is less than unity, is itself less than either of its factors. Therefore the square, cube, etc., of a fraction less than unity decrease, and the smaller the fraction is the more rapid is that decrease, as the

*When any quantity is increased, the quantity by which it is increased is called its *increment*.

following examples will show:

$$\begin{aligned} \text{Let}\quad x &= .01, \\ x^2 &= .0001, \\ x^3 &= .000001, \\ &\text{etc.} \end{aligned} \qquad \begin{aligned} \text{Let}\quad x &= .00001, \\ x^2 &= .0000000001, \\ x^3 &= .000000000000001, \\ &\text{etc.} \end{aligned}$$

Now quantities are compared, not by the actual difference which exists between them, but by the number of times which one contains the other, and, of two quantities which are both very small, one may be very great as compared with the other. In the second example x^2 and x^3 are both small fractions when compared with unity; nevertheless, x^2 is very great when compared with x^3, being 100,000 times its magnitude. This use of the words small and great sometimes embarrasses the beginner; nevertheless, on consideration, it will appear to be very similar to the sense in which they are used in common life. We do not form our ideas of smallness or greatness from the actual numbers which are contained in a collection, but from the proportion which the numbers bear to those which are usually found in similar collections. Thus of 1000 men we should say, if they lived in one village, that it was extremely large; if they formed a regiment, that it was rather large; if an army, that it was utterly insignificant in point of numbers. Hence, in such an expression as $Ah+Bh^2+Ch^3$, we may, if h is very small, reject Bh^2+Ch^3, as being very small compared with Ah. An error will thus be committed, but a very small one only, and which becomes smaller as h becomes smaller.

Let us take any algebraical expression, such as $x^2 + x$, and suppose that x is increased by a very small quantity h. The expression then becomes $(x+h)^2 + (x+h)$, or $x^2 + x + (2x+1)h + h^2$. But it was $x^2 + x$; therefore, in consequence of x receiving the increment h, $x^2 + x$ has received the increment $(2x+1)h+h^2$, for which $(2x+1)h$ may be written, since h is very small. This is proportional to h, since, if h were doubled, $(2x+1)h$ would be doubled; also, if the first were halved the second would be halved, etc. In general, if y is a quantity which contains x, and if x be changed into $x+h$, y is changed into a quantity of the form $y + Ah + Bh^2 + Ch^3 +$ etc.; that is, y receives an increment of the form $Ah + Bh^2 + Ch^3 +$ etc. If h be very small, this may, without sensible error, be reduced to its first term, viz., Ah, which is proportional to h. The general proof of this proposition belongs to a higher department of mathematics; nevertheless, the student may observe that it holds good in all the instances which occur in elementary treatises on arithmetic and algebra.

For example:

$$(x+h)^m = x^m + mx^{m-1}h + m\,\frac{m-1}{2}\,x^{m-2}h^2 + \text{etc.}$$

Here $A = mx^{m-1}$, $B = m\,\dfrac{m-1}{2}\,x^{m-2}$, etc.; and if h be very small, $(x+h)^m = x^m + mx^{m-1}h$, nearly.

Again, $e^h = 1 + h + \dfrac{h^2}{2} + \dfrac{h^3}{2\cdot 3} +$ etc. Therefore, $e^x \times e^h$

or $e^{x+h} = e^x + e^x h + \dfrac{e^x}{2} h^2 + \text{etc.}$ And if h be very small, $e^{x+h} = e^x + e^x h$, nearly.

Again, $\log(1+n') = M(n' - \frac{1}{2}n'^2 + \frac{1}{3}n'^3 - \text{etc.})$. To each side add $\log x$, recollecting that

$$\log x + \log(1+n') = \log x(1+n') = \log(x+xn'),$$

and let

$$xn' = h \quad \text{or} \quad n' = \frac{x}{h}.$$

Making these substitutions, the equation becomes

$$\log(x+h) = \log x + \frac{M}{x} h - \frac{M}{2x^2} h^2 + \text{etc.}$$

If h is very small, $\log(x+h) = \log x + \dfrac{M}{x} h$.

We can now apply this to the logarithmic example with which we commenced this subject. It appears that

$$\begin{aligned} \log 41713 &= 4.6202714, \\ \log(41713+1) &= 4.6202714 + .0000104, \\ \log(41713+2) &= 4.6202714 + .0000104 \times 2. \end{aligned}$$

From which, and the considerations above-mentioned,

$$\log(41713+h) = \log 41713 + .0000104 \times h,$$

which is extremely near the truth, even when h is a much larger number, as the tables will show. Suppose, then, that the logarithm of 41713.27 is required. Here $h = .27$. It therefore only remains to calculate $.0000104 \times .27$, and add the result, or as much of it as is contained in the first seven places of decimals, to the logarithm of 41713. This trouble is saved in the tables in the following manner. The difference of the successive logarithms is written down, with the exception of the cyphers at the beginning, in the column marked *D* or *Diff.*, under which are registered the tenths of that difference, or as much of them as is contained in the first seven decimal places, increasing the seventh figure by 1 when the eighth is equal to or greater than 5, and omitting the cyphers to save room. From this table of tenths the table of hundredth parts may be made by striking off the last figure, making the usual change in the last but one, when the last is equal to or greater than 5, and placing an additional cypher. The logarithm of 41713.27 is, therefore, obtained in the following manner:

$$\begin{aligned} \log 41713 \quad &= 4.6202714 \\ .0000104 \times .2 &= .0000021 \\ .0000104 \times .07 &= \underline{.0000007} \\ \log 41713.27 &= 4.620274. \end{aligned}$$

This, when the useless cyphers and parts of the operation are omitted, is the process given in all the books of logarithms. If the logarithm of a number containing more than seven significant figures be sought, for example 219034.717, recourse must be had to a table, in which the logarithms are carried to more than seven places of decimals. The fact is, that in the first seven places of decimals there is no difference between log 219034.7 and log 219034.717. For an excellent treatise on the practice of logarithms the reader may consult the preface to Babbage's *Table of Logarithms.**

*Copies of Babbage's *Table of Logarithms* are now scarce, and the reader may accordingly be referred to the prefaces of the treatises mentioned on page 169. The article on "Logarithms, Use of" in the *English Cyclopedia*, may also be consulted with profit.—*Ed.*

CHAPTER XII.

ON THE STUDY OF ALGEBRA.

In this chapter we shall give the student some advice as to the manner in which he should prosecute his studies in algebra. The remaining parts of this subject present a field infinite in its extent and in the variety of the applications which present themselves. By whatever name the remaining parts of the subject may be called, even though the ideas on which they are based may be geometrical, still the mechanical processes are algebraical, and present continual applications of the preceding rules and developments of the subjects already treated. This is the case in Trigonometry, the application of Algebra to Geometry, the Differential Calculus, or Fluxions, etc.

I. The first thing to be attended to in reading any algebraical treatise, is the gaining a perfect understanding of the different processes there exhibited, and of their connexion with one another. This cannot be attained by a mere reading of the book, however great the attention which may be given. It is impossible, in a mathematical work, to fill up every process in the manner in which it must be filled up in the mind of the student before he can be said to have completely mastered it. Many results must be given, of which the details are suppressed, such are the additions, multiplications, extractions of the square root, etc., with which the investigations abound. These must not be taken on trust by the student, but must be worked by his own pen, which must never be out of his hand while engaged in any algebraical process. The method which we recommend is, to write the whole of the symbolical part of each investigation, filling up the parts to which we have alluded, adding only so much verbal elucidation as is absolutely necessary to explain the connexion of the different steps, which will generally be much less than what is given in the book. This may appear an alarming labor to one who has not tried it, nevertheless we are convinced that it is by far the shortest method of proceeding, since the deliberate consideration which the act of writing forces us to give, will prevent the confusion and difficulties which cannot fail to embarrass the beginner if he attempt, by mere perusal only, to understand new reasoning expressed in new language. If, while proceeding in this manner, any difficulty should occur, it should be written at full length, and it will often happen that the misconception which occasioned the embarrassment will not stand the trial to which it is thus brought. Should there be still any matter of doubt which is not removed by attentive reconsideration, the student should proceed, first making a note of the point which he is unable to perceive. To this he should recur in his subsequent progress, whenever he arrives at anything which appears to have any affinity, however remote, to the difficulty which stopped him, and thus he will frequently find himself in a condition to decypher what formerly appeared incomprehensible. In reasoning purely geometrical, there is less necessity for committing to writing the whole detail of the arguments, since the symbolical language is more quickly understood, and the subject is in a great measure independent of the mechanism of operations; but, in the processes of algebra, there is no point on which so much depends, or on which it becomes an instructor more strongly to insist.

II. On arriving at any new rule or process, the student should work a number of examples sufficient to prove to himself that he understands and can apply the rule or process in question. Here a difficulty will occur, since there are many of these in the books, to which no examples are formally given. Nevertheless, he may choose an example for himself, and his previous knowledge will suggest some method of proving whether his result is true or not. For example, the development of $(a+x)^{\frac{7}{3}}$ will exercise him in the use of the binomial theorem; when he has obtained the series which is equivalent to $(a+x)^{\frac{7}{3}}$, let him, in the same way, develop $(a+x)^{\frac{2}{3}}$; the product of these, since $\frac{7}{3}+\frac{2}{3}=3$, ought to be the same as the development of $(a+x)^3$, or as $a^3+3a^2x+3ax^2+x^3$. He may also try whether the development of $(a+x)^{\frac{1}{2}}$ by the binomial theorem, gives the same result as is obtained by the extraction of the square root of $a+x$. Again, when any development is obtained, it should be seen whether the development possesses all the properties of the expression from which it has been derived. For example, $\dfrac{1}{1-x}$ is proved to be equivalent to the series

$$1+x+x^2+x^3+\text{ etc.,}\quad \textit{ad infinitum.}$$

This, when multiplied by $1 - x$, should give 1; when multiplied by $1 - x^2$, should give $1 + x$, because

$$\frac{1}{1-x} \times (1-x) = 1, \qquad \frac{1}{1-x} \times (1-x^2) = 1 + x, \quad \text{etc.}$$

Again,

$$a^x = 1 + x\log a + \frac{x^2(\log a)^2}{2} + \frac{x^3(\log a)^3}{2\cdot 3} + \ldots \textit{ad inf.},$$

$$a^y = 1 + y\log a + \frac{y^2(\log a)^2}{2} + \text{etc.},$$

$$a^{x+y} = 1 + (x+y)\log a + \frac{(x+y)^2(\log a)^2}{2} + \text{etc.}$$

Now, since $a^x \times a^y = a^{x+y}$, the product of the two first series should give the third. Many other instances of the same sort will suggest themselves, and a careful attention to them will confirm the demonstration of the several theorems, which, to a beginner, is often doubtful, on account of the generality of the reasoning.

III. Whenever a demonstration appears perplexed, on account of the number and generality of the symbols, let some particular case be chosen, and let the same demonstration be applied. For example, if the binomial theorem should not appear sufficiently plain, the same reasoning may be applied to the expansion of $(1+x)^{\frac{2}{3}}$, or any other case, which is there applied to $(1+x)^{\frac{m}{n}}$. Again, the general form of the product $(x+a)$, $(x+b)$, $(x+c)$, etc., ... containing n factors, will be made apparent by taking first two, then three, and four factors, before attempting to apply the reasoning which establishes the form of the general product. The same applies particularly to the theory of permutations and combinations, and to the doctrine of probabilities, which is so materially connected with it. In the theory of equations it will be advisable at first, instead of taking the general equation of the form

$$x^n + Ax^{n-1} + Bx^{n-2} + \cdots + Lx + M = 0,$$

to choose that of the third, or at most of the fourth degree, or both, on which to demonstrate all the properties of expressions of this description. But in all these cases, when the particular instances have been treated, the general case should not be neglected, since the power of reasoning upon expressions such as the one just given, in which all the terms cannot be written down, on account of their indeterminate number, must be exercised, before the student can proceed with any prospect of success to the higher branches of mathematics.

IV. When any previous theorem is referred to, the reference should be made, and the student should satisfy himself that he has not forgotten its demonstration. If he finds that he has done so, he should not grudge the time necessary for its recovery. By so doing, he will avoid the necessity of reading over the subject again, and will obtain the additional advantage of being able to give to each part of the subject a time nearly proportional to its importance, whereas, by reading a book over and over again until he is a master of it, he will not collect the more prominent parts, and will waste time upon unimportant details, from which even the best books are not free. The necessity for this continual reference is particularly felt in the Elements of Geometry, where allusion is constantly made to preceding propositions, and where many theorems are of no importance, considered as results, and are merely established in order to serve as the basis of future propositions.

V. The student should not lose any opportunity of exercising himself in numerical calculation, and particularly in the use of the logarithmic tables. His power of applying mathematics to questions of practical utility is in direct proportion to the facility which he possesses in computation. Though it is in plane and spherical trigonometry that the most direct numerical applications present themselves, nevertheless the elementary parts of algebra abound with useful practical questions. Such will be found resulting from the binomial theorem, the theory of logarithms, and that of continued fractions. The first requisite in this branch of the subject, is a perfect acquaintance with the arithmetic of decimal fractions; such a degree of acquaintance as can only be gained by a knowledge of the principles as well as of the rules which are deduced from them. From the imperfect manner in which arithmetic is usually taught, the student ought in most cases to recommence this study before proceeding to the practice of logarithms.

VI. The greatest difficulty, in fact almost the only one of any importance which algebra offers to the reason, is the use of the isolated negative sign in such expressions as $-a$, a^{-x}, and the symbols which we have called imaginary. It is a remarkable fact, that the first elements of the mathematics, sciences which demonstrate their results with more certainty than any others, contain difficulties which have been the subjects of discussion for centuries. In geometry, for example, the theory of parallel lines has never yet been freed from the difficulty which presented itself to Euclid, and obliged him to assume, instead of proving, the 12th axiom of his first book. Innumerable as have been the attempts to elude or surmount this obstacle, no one has been more successful than another. The elements of fluxions or the differential calculus, of mechanics, of optics, and of all the other sciences, in the same manner contain difficulties peculiar to themselves. These are not such as would suggest themselves to the beginner, who is usually embarrassed by the actual performance of the operations, and no ways perplexed by any doubts as to the foundations of the rules by which he is to work. It is the characteristic of a young student in the mathematical sciences, that he sees, or fancies that he sees, the truth of every result which can be stated in a few words, or arrived at by few and simple operations, while that which is long is always considered by him as abstruse. Thus while he feels no embarrassment as to the meaning of the equation $+a \times -a = -a^2$, he considers the multiplication of $a^m + a^n$ by $b^m + b^n$ as one of the difficulties of algebra. This arises, in our opinion, from the manner in which his previous studies are usually conducted. From his earliest infancy, he learns no fact from his own observation, he deduces no truth by the exercise of his own reason. Even the tables of arithmetic, which, with a little thought and calculation, he might construct for himself, are presented to him ready made, and it is considered sufficient to commit them to memory. Thus a habit of examination is not formed, and the student comes to the science of algebra fully prepared to believe in the truth of any rule which is set before him, without other authority than the fact of finding it in the book to which he is recommended. It is no wonder, then, that he considers the difficulty of a process as proportional to that of remembering and applying the rule which is given, without taking into consideration the nature of the reasoning on which the rule was founded. We are not advocates for stopping the progress of the student by entering fully into all the arguments for and against such questions, as the use of negative quantities, etc., which he could not understand, and which are inconclusive on both sides; but he might be made aware that a difficulty does exist, the nature of which might be pointed out to him, and he might then, by the consideration of a sufficient number of examples, treated separately, acquire confidence in the results to which the rules lead. Whatever may be thought of this method, it must be better than an unsupported rule, such as is given in many works on algebra.

It may perhaps be objected that this is induction, a species of reasoning which is foreign to the usually received notions of mathematics. To this it may be answered, that inductive reasoning is of as frequent occurrence in the sciences as any other. It is certain that most great discoveries have been made by means of it; and the mathematician knows that one of his most powerful engines of demonstration is that peculiar species of induction which proves many general truths by demonstrating that, if the theorem be true in one case, it is true for the succeeding one. But the beginner is obliged to content himself with a less rigorous species of proof, though equally conclusive, as far as moral certainty is concerned. Unable to grasp the generalisations with which the more advanced student is familiar, he must satisfy himself of the truth of general theorems by observing a number of particular simple instances which he is able to comprehend. For example, we would ask any one who has gone over this ground, whether he derived more certainty as to the truth of the binomial theorem from the general demonstration (if indeed he was suffered to see it so early in his career), or from observation of its truth in the particular cases of the development of $(a + b)^2$, $(a + b)^3$, etc., substantiated by ordinary multiplication. We believe firmly, that to the mass of young students, general demonstrations afford no conviction whatever; and that the same may be said of almost every species of mathematical reasoning, when it is entirely new. We have before observed, that it is necessary to learn to reason; and in no case is the assertion more completely verified than in the study of algebra. It was probably the experience of the inutility of general demonstrations to the very young student that caused the abandonment of reasoning which prevailed so much in English works on elementary mathematics. Rules which the student could follow in practice supplied the place of arguments which he could not, and no pains appear to have been taken to adopt a middle course, by suiting the nature of the proof to the student's capacity. The objection to this appears to have been the necessity which arose for departing from the appearance of rigorous demonstration. This was the cry of those who, not having seized the spirit of the processes which they followed, placed the force of the reasoning in the forms. To such the authority of great names is a strong argument; we will therefore cite the words of Laplace on this subject.

"Newton extended to fractional and negative powers the analytical expression which he had found for whole and positive ones. You see in this extension one of the great advantages of algebraic language which expresses truths much more general than those which were at first contemplated, so that by making the extension of which it admits, there arises a multitude of new truths out of formulæ which were founded upon very limited suppositions. At first, people were afraid to admit the general consequences with which analytical formulæ furnished them; *but a great number of examples having verified them*, we now, without fear, yield ourselves to the guidance of analysis through all the consequences to which it leads us, and the most happy discoveries have sprung from the boldness. We must observe, however, that precautions should be taken to avoid giving to formulæ a greater extension than they really admit, and that it is always well to demonstrate rigorously the results which are obtained."

We have observed that beginners are not disposed to quarrel with a rule which is easy in practice, and verified by examples, on account of difficulties which occur in its establishment. The early history of the sciences presents occasion for the same remark. In the work of Diophantus, the first Greek writer on algebra, we find a principle equivalent to the equations $+a \times -b = -ab$, and $-a \times -b = +ab$, admitted as an axiom, without proof or difficulty. In the Hindoo works on algebra, and the Persian commentators upon them, the same thing takes place. It appears, that struck with the practical utility of the rule, and certain by induction of its truth, they did not scruple to avail themselves of it. A more cultivated age, possessed of many formulæ whose developments presented striking examples of an universality in algebraic language not contemplated by its framers, set itself to inquire more closely into the first principles of the science. Long and still unfinished discussions have been the result, but the progress of nations has exhibited throughout a strong resemblance to that of individuals.

VII. The student should make for himself a syllabus of results only, unaccompanied by any demonstration. It is essential to acquire a correct memory for algebraical formulæ, which will save much time and labor in the higher departments of the science. Such a syllabus will be a great assistance in this respect, and care should be taken that it contain only the most useful and most prominent formulæ. Whenever that can be done, the student should have recourse to the system of tabulation, of which he will have seen several examples in this treatise. In this way he should write the various forms which the roots of the equation $ax^2 + bx + c = 0$ assume, according to the signs of a, b, and c, etc. Both the preceptor and the pupil, but especially the former, will derive great advantage from the perusal of Lacroix, *Essais sur l'Enseignement en général et sur celui des Mathématiques en particulier*,* Condillac, *La Langue des Calculs*, and the var-

*The books mentioned in the present passage, while still very valuable, are now not easily procurable and, besides, do not give a complete idea of the subject in its modern extent. A recent work on the *Philosophy and Teaching of Mathematics* is that of C. A. Laisant (*La Mathématique. Philosophie-Enseignement*, Paris, 1898, Georges Carré et C. Naud, publishers.) Perhaps the most accessible and useful work in English for the elements is David Eugene Smith's new book *The Teaching of Elementary Mathematics.* (New York: The Macmillan Company, 1900). Mention might be made also of W. M. Gillespie's translation from Comte's *Cours de Philosophie Positive*, under the title of *The Philosophy of Mathematics* (New York: Harpers, 1851), and of the *Cours de Méthodologie Mathématique* of Félix Dauge (Deuxième edition, revue et augmentee. Gand, Ad. Hoste; Paris, Gauthier-Villars, 1896). The recent work of Freycinet on the *Philosophy of the Sciences* (Paris, 1896, Gauthier-Villars) will be found valuable. One of the best and most comprehensive of the modern works is that of Duhamel, *Des Méthodes dans les Sciences de Raisonnement*, (5 parts, Paris, Gauthier-Villars), a work giving a comprehensive exposition of the foundations of all the mathematical sciences. The chapters in Dühring's *Kritische Geschichte der Prinzipien der Mechanik* and his *Neue Grundmittel* on the study of mathematics and mechanics is replete with original, but hazardous, advice, and may be consulted as a counter-irritant to the traditional professional views of the subject. The articles in the *En-*

ious articles on the elements of algebra in the French Encyclopedia, which are for the most part written by D'Alembert. The reader will here find the first principles of algebra, developed and elucidated in a masterly manner. A great collection of examples will be found in most elementary works, but particularly in Hirsch, *Sammlung von Beispielen*, etc., translated into English under the title of *Self-Examinations in Algebra*, etc., London: Black, Young and Young, 1825.* The student who desires to carry his algebraical studies farther than usual, and to make them the stepping-stone to a knowledge of the higher mathematics, should be acquainted with the French language.† A knowledge of this, sufficient to enable him to read the simple and easy style in which the writers of that nation treat the first principles of every subject, may be acquired in a short time. When that is done, we recommend to the student the algebra of M. Bourdon,‡

---

glish Cyclopedia, by De Morgan himself, contain refreshing hints on this subject. But the greatest inspiration is to be drawn from the works of the masters themselves; for example, from such works as Laplace's Introduction to the *Calculus of Probabilities*, or from the historical and philosophical reflexions that uniformly accompany the later works of Lagrange. The same remark applies to the later mathematicians of note.—*Ed.*

*Hirsch's *Collection*, enlarged and modernised, can be obtained in various recent German editions. The old English translations of the original are not easily procured.—*Ed.*

†German is now of as much importance as French. But the French text-books still retain their high standard.—*Ed.*

‡Bourdon's *Elements of Algebra* is still used in France, having ap-

a work of eminent merit, though of some difficulty to the English student, and requiring some previous habits of algebraical reasoning.

VIII. The height to which algebraical studies should be carried, must depend upon the purpose to which they are

---

peared in 1895 in its eighteenth edition, with notes by M. Prouhet (Gauthier-Villars, Paris.) A more elementary French work of a modern character is that of J. Collin (Second edition, 1888, Paris, Gauthier-Villars). A larger and more complete treatise which begins with the elements and extends to the higher branches of the subject is the *Traité d'Algèbre* of H. Laurent, in four small volumes (Gauthier-Villars, Paris). This work contains a large collection of examples. Another elementary work is that of C. Bourlet, *Lecons d'Algèbre Elémentaire*, Paris, Colin, 1896. A standard and exhaustive work on higher algebra is the *Cours d'Algèbre Supérieure*, of J. A. Serret, two large volumes (Fifth edition, 1885, Paris, Gauthier-Villars).

The number of American and English text-books of the intermediate and higher type is very large. Todhunter's *Algebra and Theory of Equations* (London: Macmillan & Co.) were for a long time the standards in England and this country, but have now (especially the first-mentioned) been virtually superseded. An excellent recent text-book for beginners, and one that skilfully introduces modern notions, is the *Elements of Algebra* of W. W. Beman and D. E. Smith (Boston, 1900). Fisher and Schwatt's elementary text-books of algebra are also recommendable from both a practical and theoretical point of view. Valuable are C. Smith's *Treatise on Algebra* (London: Macmillan), and Oliver, Wait, and Jones's *Treatise on Algebra* (Ithaca, N. Y., 1887), also Fine's *Number System of Algebra* (Boston: Leach). The best English work on the theory of equations is Burnside and Panton's (Longmans).

A very exhaustive presentation of the subject from the modern point of view is the *Algebra* of Professor George Chrystal (Edinburgh: Adam and Charles Black, publishers), in two large volumes of nearly six hunto be applied. For the ordinary purposes of practical mathematics, algebra is principally useful as the guide to trigonometry, logarithms, and the solution of equations. Much and profound study is not therefore requisite; the student should pay great attention to all numerical processes and partic-

---

dred pages each. Recently Professor Chrystal has published a more elementary work entitled *Introduction to Algebra* (same publishers).

A few German works may also be mentioned in this connexion, for the benefit of readers acquainted with that language. Professor Hermann Schubert has, in various forms, given systematic expositions of the elementary principles of arithmetic, (e. g., see his *Arithmetik und Algebra*, Sammlung Göschen, Leipsic,—an extremely cheap series containing several other elementary mathematical works of high standard; also, for a statement of Schubert's views in English consult his *Mathematical Recreations*, Chicago, 1898). Professor Schubert has recently begun the editing of a new and larger series of mathematical text-books called the *Sammlung Schubert* (Leipsic: Göschen), which contains three works treating of algebra. In this connexion may be mentioned also Matthiessen's admirable *Grundzüge der antiken und modernen Algebra* (Leipsic: Teubner) for literal equations. The following are all excellent: (1) Otto Biermann's *Elemente der höheren Mathematik* (Leipsic, 1895); (2) Petersen's *Theorie der algebraischen Gleichungen* (Copenhagen: Höst; also in French, Paris: Gauthier-Villars); (3) Richard Baltzer's *Elemente der Mathematik* (2 vols., Leipsic: Hirzel); (4) Gustav Holzmüller's *Methodisches Lehrbuch der Elementarmathematik* (3 parts, Leipsic: Teubner); (5) Werner Jos. Schüller's *Arithmetik und Algebra für höhere Schulen und Lehrerseminare, besonders zum Selbstunterricht*, etc. (Leipsic, 1891, Teubner); (6) Oskar Schlömilch's *Handbuch der algebraischen Analysis* (Frommann, Stuttgart); (7) Eugen Netto's *Vorlesungen über Algebra* (Leipsic: Teubner, 2 vols.); (8) Heinrich Weber's *Lehrbuch der Algebra* (Braunschweig: Vieweg, 2 vols). This last work is the most advanced treatise

ularly to the methods of approximation which he will find in all the books. His principal instrument is the table of logarithms of which he should secure a knowledge both theoretical and practical. The course which should be adopted preparatory to proceeding to the higher branches of mathematics is different. It is still of great importance that the student should be well acquainted with numerical applications; nevertheless, he may omit with advantage many details relative to the obtaining of approximative numerical results, particularly in the theory of equations of higher degrees than the second. Instead of occupying himself upon these, he should proceed to the application of algebra to geometry, and afterwards to the differential calculus. When a competent knowledge of these has been obtained, he may then revert to the subjects which he has neglected, giving them more or less attention according to his own opinion of the use which he is likely to have for them. This applies particularly to the theory of equations, which abounds with processes of which very few students will afterwards find the necessity.

We shall proceed in the next number to the difficulties which arise in the study of Geometry and Trigonometry.

---

that has yet appeared. A French translation has been announced.—*Ed.*—April, 1902.

CHAPTER XIII.

ON THE DEFINITIONS OF GEOMETRY.

In this treatise on the difficulties of Geometry and Trigonometry, we propose, as in the former part of the work, to touch on those points only which, from novelty in their principle, are found to present difficulties to the student, and which are frequently not sufficiently dwelt upon in elementary works. Perhaps it may be asserted, that there are no difficulties in geometry which are likely to place a serious obstacle in the way of an intelligent beginner, except the temporary embarrassment which always attends the commencement of a new study; that, for example, there is nothing in the elements of pure geometry comparable, in point of complexity, to the theory of the negative sign, of fractional indices, or of the decomposition of an expression of the second degree into factors. This may be true; and were it only necessary to study the elements of this science for themselves, without reference to their application, by means of algebra, to higher branches of knowledge, we should not have thought it necessary to call the attention of our readers to the points which we shall proceed to place before them. But while there is a higher study in which elementary ideas, simple enough in their first form, are so generalised as to become difficult, it will be an assistance to the beginner who intends to proceed through a wider course of pure mathematics than forms part of common education, if his attention is early directed, in a manner which he can comprehend, to future extensions of what is before him.

The reason why geometry is not so difficult as algebra, is to be found in the less general nature of the symbols employed. In algebra a general proposition respecting numbers is to be proved. Letters are taken which may represent any of the numbers in question, and the course of the demonstration, far from making any use of a particular case, does not even allow that any reasoning, however general in its nature, is conclusive, unless the symbols are as general as the arguments. We do not say that it would be contrary to good logic to form general conclusions from reasoning on one particular case, when it is evident that the same considerations might be applied to any other, but only that very great caution, more than a beginner can see the value of, would be requisite in deducing the conclusion. There occurs also a mixture of general and particular propositions, and the latter are liable to be mistaken for the former. In geometry on the contrary, at least in the elementary parts, any proposition may be safely demonstrated by reasonings on any one particular example. For though in proving a property of a triangle many truths regarding that triangle may be asserted as having been proved before, none are brought forward which are not general, that is, true for all instances of the same kind. It also affords some facility that the results of elementary geometry are in many cases sufficiently evident of themselves to the eye; for instance, that two sides of a triangle are greater than the third, whereas in algebra many rudimentary propositions derive no evidence from the senses; for example, that $a^3 - b^3$ is always divisible without remainder by $a - b$.

The definitions of the simple terms *point*, *line*, and *surface* have given rise to much discussion. But the difficulties which attend them are not of a nature to embarrass the beginner, provided he will rest content with the notions which he has already derived from observation. No explanation can make these terms more intelligible. To them may be added the words *straight line*, which cannot be mistaken for one moment, unless it be by means of the attempt to explain them by saying that a straight line is "that which lies evenly between its extreme points."

The line and surface are distinct species of magnitude, as much so as the yard and the acre. The first is no part of the second, that is, no number of lines can make a surface. When therefore a surface is divided into two parts by a line, the dividing line is not to be considered as forming a part of either. That the idea of the line or boundary necessarily enters into the notion of the division is very true; but if we conceive the line abstracted, and thus get rid of the idea of division, neither surface is increased or diminished, which is what we mean when we say that the line is not a part of the surface. The same considerations apply to a point, considered as the boundary of the divisions of a line.

The beginner may perhaps imagine that a line is made up of points, that is, that every line is the sum of a number of points, a surface the sum of a number of lines, and so on. This arises from the fact, that the things which we draw on paper as the representatives of lines and points, have in reality three dimensions, two of which, length and breadth, are perfectly visible. Thus the point, such as we are obliged to represent it, in order to make its position visible, is in reality a part of our line, and our points, if sufficiently multiplied in number and placed side by side, would compose a line of any length whatever. But taking the mathematical definition of a point, which denies it all magnitude, either in length, breadth, or thickness, and of a line, which is asserted to possess length only without breadth or thickness, it is easy to show that a point is no part of a line, by making it appear that the shortest line can be cut in as many points as the longest, which may be done in the following manner. Let AB be any straight line, from the ends of which, A and B, draw two lines, AF and CB, parallel to one another. Consider AF as produced without limit, and in CB take any point C, from which draw lines CE, CF, etc., to different points in AF. It is evident that for each point E in AF there is a distinct point in AB, viz., the intersection of CE with AB;—for, were it possible that two points, E and F in AF, could be thus connected with the same point of AB, it is evident that two straight lines would enclose a space, viz., the lines CE and CF, which both pass through C, and would, were our supposition correct, also pass through the same point in AB. There can then be taken as many points in the finite or unbounded line AB as in the indefinitely extended line AF.

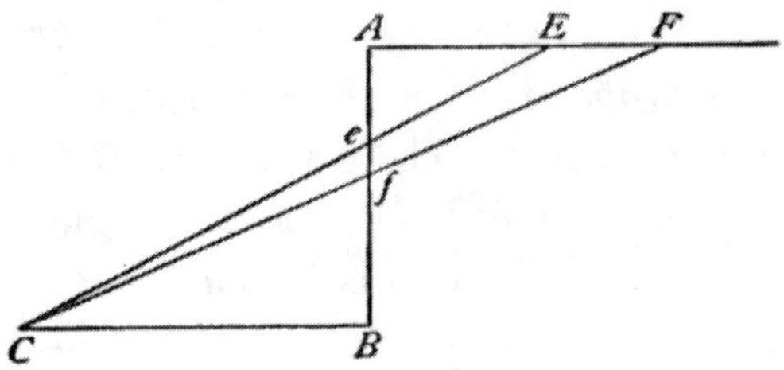

Fig. 1.

The next definition which we shall consider is that of a *plane surface.* The word *plane* or *flat* is as hard to define, without reference to any thing but the idea we have of it, as it is easy to understand. Nevertheless the practical method of ascertaining whether or no a surface is plane, will furnish a definition, not such, indeed, as to render the nature of a plane surface more evident, but which will serve, in a mathematical point of view, as a basis on which to rest the propositions of solid geometry. If the edge of a ruler, known to be perfectly straight, coincides with a surface throughout its whole length, in whatever direction it may be placed upon that surface, we conclude that the surface is plane. Hence the definition of a plane surface is that in which, any two points being taken, the straight line joining these points lies wholly upon the surface.

Two straight lines have a relation to one another independent altogether of their length. This we commonly express (for among the most common ideas are found the germ of every geometrical theory) by saying that they are in the same or different *directions.* By the *direction* of the needle we ascertain the *direction* in which to proceed at sea, and by the *direction* in which the hands of a clock are placed we tell the hour. It remains to reduce this common notion to a more precise form.

Suppose a straight line OA to be given in magnitude and position, and to remain fixed while another line OB, at first coincident with OA, is made to move round OA, so as continually to vary its direction with respect to OA. The process of opening a pair of compasses will furnish an illustration of this, but the two lines need not be equal to one another. In this case the opening made by the two will continually increase, and this opening is a species of magnitude, since one opening may be compared with another, so as to ascertain which of the two is the greater. Thus if the figure CPD be removed from its place, without any other change, so that the point P may fall on O, and the line PC may lie upon and become a part of OA, or OA of PC, according to which is the longer of the two, then if the opening CPD is the same as the opening AOB, PD will lie upon OB at the same time as PC lies upon OA. But if PD does not then lie upon OB, but falls between OB and OA, the opening CPD is less than the opening AOB, and if PD does not fall between OA and OB, or on OB, the opening CPD is greater than the opening BOA. To this species of magnitude, the opening of two lines, the name of angle is given, that is BO is said to make an angle with OA. The difficulty here arises

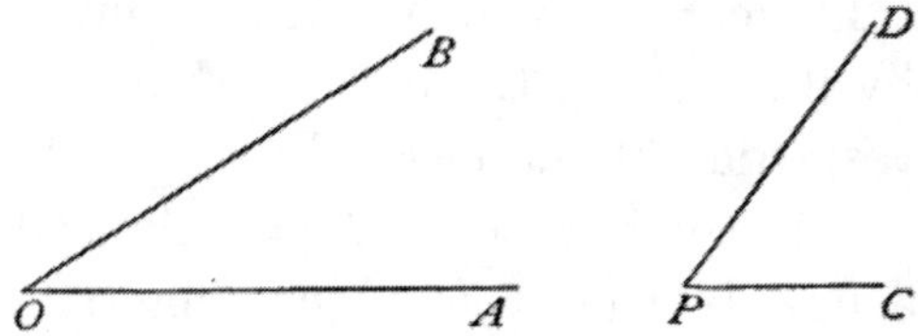

Fig. 2.

from this magnitude being one, the measure of which has seldom fallen under observation of those who begin geometry. Every one has measured one line by means of another, and has thus made a number the representative of a length; but few, at this period of their studies, have been accustomed to the consideration, that one opening may be contained a certain number of times in another, or may be a certain fraction of another. Nevertheless we may find measures of this new species of magnitude either by means of time, length, or number.

One magnitude is said to be a measure of another, when, if the first be doubled, trebled, halved, etc., the second is doubled, trebled, or halved, etc.; that is, when any fraction or multiple of the first corresponds to the same fraction or multiple of the second in the same manner as the first does to the second. The two quantities need not be of the same kind: thus, in the barometer the height of the mercury (a length) measures the pressure of the atmosphere (a weight); for if the barometer which yesterday stood at 28 inches, to-day stands at 29 inches, in which case the height of yesterday is increased by its 28th part, we know that the atmospheric pressure of yesterday is increased by its 28th part to-day. Again, in a watch, the *number of hours* elapsed since twelve o clock is measured by the *angle* which a hand makes with the position it occupied at twelve o'clock. In the spring balances a *weight* is measured by an *angle*, and many other similar instances might be given.

This being premised, suppose a line which moves round another as just described, to move uniformly, that is, to describe equal openings or angles in equal times. Suppose the line OA to move completely round, so as to reassume its first position in twenty-four hours. Then in twelve hours the moving line will be in the position OB, in six hours it will be in OC, and in eighteen hours in OD. The line OC is that which makes equal angles with OA and OB, and is said to be at right angles, or perpendicular to OA and OB. Again, OA and OB which are in the same right line, but on opposite sides of the point O, evidently make an opening or angle which is equal to the sum of the angles AOC and COB, or equal to two right angles. A line may also be said to make with itself an opening equal to four right angles, since after revolving through four right angles, the moving line reassumes its original position. We may even carry this notion farther: for if the moving line be in the position OE when P hours have elapsed, it will recover that position after every twenty-four hours, that is, for every additional four right angles described; so that the angle AOE is equally well

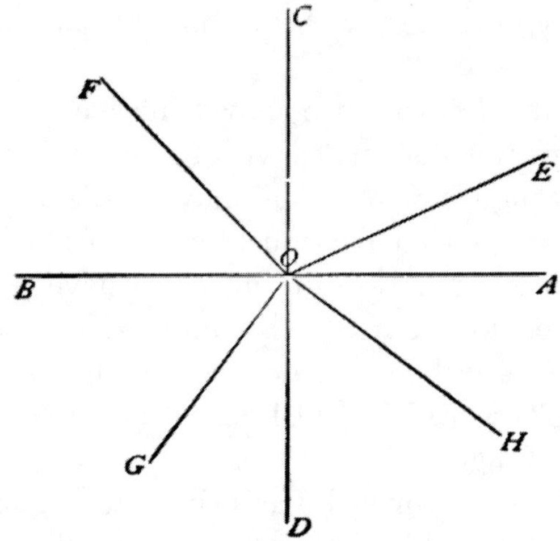

Fig. 3.

represented by any of the following angles:

$$4 \text{ right angles} + AOE,$$
$$8 \text{ right angles} + AOE,$$
$$12 \text{ right angles} + AOE, \quad \text{etc.}$$

These formulæ which suppose an opening greater than any *apparent* opening, and which take in and represent the fact that the moving line has attained its position for the second, third, fourth, etc., time, since the commencement of the motion, are not of any use in elementary geometry; but as they play an important part in the application of algebra to the theory of angles, we have thought it right to mention them here.

It is plain also that we may conceive the line OE to make two openings or angles with the original position OA: (1) that through which it has moved to recede from OA; (2) that through which it must move to reach OA again. The first (in the position in which we have placed OA) is what is called in geometry the angle AOE; the second is more simply described as composed of the openings or angles EOC, COB, BOD, DOA, and is not used except in the application of algebra above mentioned.* Of the two angles just alluded to, one must be less than two right angles, and the second greater; the first is the one usually referred to.

It is plain that the angle or opening made by two lines does not depend upon their length but upon their position; if either be shortened or lengthened, the angle still remains the same; and if while the angle increases or decreases one of the straight lines containing it is diminished, the angle so contained may have a definite and given magnitude at the moment when the straight line disappears altogether and becomes nothing. For example, take two points of any curve AB, and join A and B by a straight line. Let the point B move towards A; it is evident that the angle made by the moving line with AB increases continually, while as much of one of the lines containing it as is intercepted by the

*But use is made of it in some modern text-books of elementary geometry.—*Ed.*

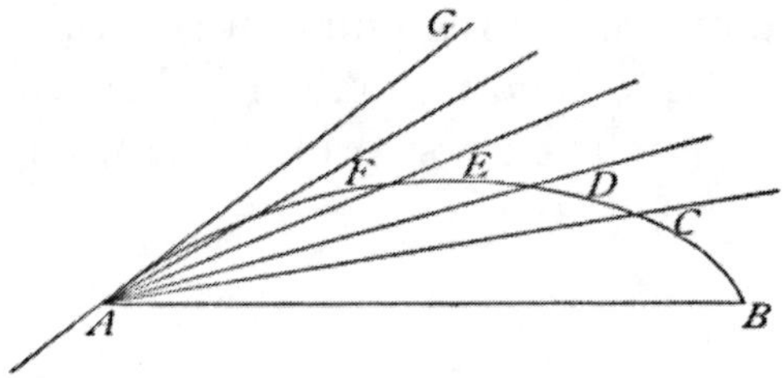

Fig. 4.

curve, diminishes without limit. When this intercepted part disappears entirely, the line in which it would have lain had it had any length, has reached the line AG, which is called the tangent of the curve.

In elementary geometry two equal angles lying on different sides of a line, such as AOE, AOH (Fig. 3), would be considered as the same. In the application of algebra, they would be considered as having different signs, for reasons stated at length in pages 111 et seq., of the first part of this Treatise. It is also common in the latter branch of the science to measure angles in one direction only; for example, in Figure 3 the angles made by OE, OF, OG, and OH, if measured upwards from OA, would be the openings through which a line must move *in the same direction* from OA, to attain those positions; and the second, third, and fourth angles would be greater than one, two, and three right angles respectively.

We proceed to the method of reasoning in geometry, or

rather to the method of reasoning in general, since there is, or ought to be, no essential difference between the manner of deducing results from first principles, in any science.

CHAPTER XIV.

ON GEOMETRICAL REASONING.

It is evident that all reasoning, of what form soever, can be reduced at last to a number of simple propositions or assertions; each of which, if it be not self-evident, depends upon those which have preceded it. Every assertion can be divided into three distinct parts. Thus the phrase, "all right angles are equal," consists of: (1) the *subject* spoken of, viz., right angles, which is here spoken of universally, since every right angle is a part of the subject; (2) the *copula*, or manner in which the two are joined together, which is generally the verb *is*, or *is equal to*, and can always be reduced to one or the other: in this case the copula is affirmative; (3) the *predicate*, or thing asserted of the subject, viz., equal angles. The phrase, thus divided, stands as written below under 1, and is called a *universal affirmative.* The second is called a *particular affirmative* proposition; the third a *universal negative*; the fourth a *particular negative*:

1. All right angles are equal (magnitudes).
2. Some triangles are equilateral (figures).
3. No circle is convex to its diameter.
4. Some triangles are not equilateral (figures).

Many assertions appear in a form which, at first sight, cannot be reduced to one of the preceding; the following are instances of the change which it is necessary to make in them:

1. Parallel lines never meet, or parallel lines are lines

which never meet.

2. The angles at the base of an isosceles triangle are equal, or an isosceles triangle is a triangle having the angles at the base equal.

The different species of assertions, and the arguments which are compounded of them, may be distinctly conceived by referring them all to one species of subject and predicate. Since every assertion, generally speaking, includes a number of individual cases in its subject, let the points of a circle be the subject and those of a triangle the predicate. These figures being drawn, the four species of assertions just alluded to are as follows:

1. Every point of the circle is a point of the triangle, or the circle is contained in the triangle.

2. Some points of the circle are points of the triangle, or part of the circle is contained in the triangle.

3. No point of the circle is a point of the triangle, or the circle is entirely without the triangle.

4. Some points of the circle are not points of the triangle, or part of the circle is outside the triangle.

On these we observe that the second follows from the first, as also the fourth from the third, since that which is true of all is true of some or any; while the first and third do not follow from the second and fourth, *necessarily*, since that which is true of some only need not be true of all. Again, the second and fourth are not necessarily inconsistent with each other for the same reason. Also two of these assertions must be true and the others untrue. The first and the third are called *contraries*, while the first and fourth, and the second and third are *contradictory*. The *converse* of a proposition is made by changing the predicate into the subject, and the subject into the predicate. No mistake is more common than confounding together a proposition and its converse, the tendency to which is rather increased in those who begin geometry, by the number of propositions which they find, the converses of which are true. Thus all the definitions are necessarily conversely true, since the identity of the subject and predicate is not merely asserted, but the subject is declared to be a name *given* to *all* those magnitudes which have the properties laid down in the predicate, and to no others. Thus a square is a four-sided figure having equal sides and one right angle, that is, let every four-sided figure having, etc., *be called* a square, and let no other figure be called by that name, whence the truth of the converse is evident. Also many of the facts proved in geometry are conversely true. Thus all equilateral triangles are equiangular, from which it is proved that *all* equiangular triangles are equilateral. Of the first species of assertion, the universal affirmative, the converse is not necessarily true. Thus "every point in figure A is a point of B," does not imply that "every point of B is a point of A," although this may be the case, and is, if the two figures coincide entirely. The second species, the particular affirmative, is conversely true, since if some points of A are points of B, some points of B are also points of A. The first species of assertion is conversely true, if the converse be made to take the form of the second species: thus from "all right angles are equal," it may be inferred that "*some* equal magnitudes are right angles." The third species, the universal negative, is conversely true, since if "no point of B is a point of A," it may be inferred that "no point of A is a point of B." The fourth species, the particular negative, is not necessarily conversely true. From "some points of A are not points of B," or A is not entirely contained within B, we can infer nothing as to whether B is or is not entirely contained in A. It is plain that the converse of a proposition is not necessarily true, if it says more either of the subject or predicate than was said before. Now "every equilateral triangle is equiangular," does not speak of all equiangular triangles, but asserts that *among* all possible equiangular triangles are to be found *all* the equilateral ones. There may then, for anything to the contrary to be discovered in our assertion, be classes of equiangular triangles not included under this assertion, of which we can therefore say nothing. But in saying "no right angles are unequal," that which we exclude, we exclude from all unequal angles, and therefore "no unequal angles are right angles" is not more general than the first.

The various assertions brought forward in a geometrical demonstration must be derived in one of the following ways:

I. *From definition.* This is merely substituting, instead of a description, the name which it has been agreed to give to whatever bears that description. No definition ought to be introduced until it is certain that the thing defined is really possible. Thus though parallel lines are defined to be "lines which are in the same plane, and which being ever so far produced never meet," the mere agreement to call such lines, should they exist, by the name of parallels, is no sufficient ground to assume that they do exist. The definition is therefore inadmissible until it is really shown that there are such things as lines which being in the same plane never meet. Again, before applying the name, care must be taken that all the circumstances connected with the definition have been attended to. Thus, though in plane geometry, where all lines are in one plane, it is sufficient that two lines would never meet though ever so far produced, to call them parallel, yet in solid geometry the first circumstance must be attended to, and it must be shown that lines are in the same plane before the name can be applied. Some of the axioms come so near to definitions in their nature, that their place may be considered as doubtful. Such are, "the whole is greater than its part," and "magnitudes which entirely coincide are equal to one another."

II. *From hypothesis.* In the statement of every proposition, certain connexions are supposed to exist from which it is asserted that certain consequences will follow. Thus "in an isosceles triangle the angles at the base are equal," or, "if a triangle be isosceles the angles at the base will be equal." Here the hypothesis or supposition is that the triangle has two equal sides, the consequence asserted is that the angles at the base or third side will be equal. The consequence being only asserted to be true when the angle is isosceles, such a triangle is supposed to be taken as the basis of the reasonings, and the condition that its two sides are equal, when introduced in the proof, is said to be introduced by hypothesis.

In order to establish the result it may be necessary to draw other lines, etc., which are not mentioned in the first hypothesis. These, when introduced, form what is called the construction.

There is another species of hypothesis much in use, principally when it is required to deduce the converse of a theorem from the theorem itself. Instead of proving the consequence directly, the contradictory of the consequence is assumed to hold good, and if from this new hypothesis, supposed to exist together with the old one, any evidently absurd result can be derived, such as that the whole is greater than its part, this shows that the two hypotheses are not consistent, and that if the first be true, the second cannot be so. But if the second be not true, its contradictory is true, which is what was required to be proved.

III. *From the evidence of the assertions themselves.* The propositions thus introduced without proof are only such as are in their nature too simple to admit of it. They are called axioms. But it is necessary to observe, that the claim of an assertion to be called an axiom does not depend only on its being self-evident. Were this the case many propositions which are always proved might be assumed; for example, that two sides of a triangle are greater than the third, or that a straight line is the shortest distance between two points. In addition to being self-evident, it must be incapable of proof by any other means, and it is one of the objects of geometry to reduce the demonstrations to the least possible number of axioms. There are only two axioms which are distinctly geometrical in their nature, viz., "two straight lines cannot enclose a space," and "through each point outside a line, not more than one parallel to that line can be drawn." All the rest of the propositions commonly given as axioms are either arithmetical in their nature; such as "the whole is greater than its part," "the doubles of equals are equals," etc.; or mere definitions, such as "magnitudes which entirely coincide are equal;" or theorems admitting of proof, such as "all right angles are equal." There is however one more species of self-evident proposition, the postulate or self-evident problem, such as the possibility of drawing a right line, etc.

IV. *From proof already given.* What has been proved once may be always taken for granted afterwards. It is evident that this is merely for the sake of brevity, since it would be possible to begin from the axioms and proceed direct to the proof of any one proposition, however far removed from them; and this is an exercise which we recommend to the student. Thus much for the legitimate use of any single assertion or proposition. We proceed to the manner of deducing a third proposition from two others.

It is evident that no assertion can be the direct and necessary consequence of two others, unless those two contain something in common, or which is spoken of in both. In many, nay most, cases of ordinary conversation and writing, we leave out one of the assertions, which is, usually speaking, very evident, and make the other assertion followed by the consequence of both. Thus, "Geometry is useful, and therefore ought to be studied," contains not only what is expressed, but also the following, "That which is useful ought to be studied;" for were this not admitted, the former assertion would not be necessarily true. This may be written thus:

> Every thing useful is what ought to be studied.
>
> Geometry is useful, therefore geometry is what ought to be studied.

This, in its present state, is called a syllogism, and may be compared with the following, from which it only differs in the *things* spoken of, and not in the *manner* in which they are spoken of.

> Every point of the circle is a point of the triangle.
>
> The point B is a point of the circle.
>
> Therefore the point B is a point of the triangle.

Here a connexion is established between the point B and the points of the triangle (viz., that the first is one of the second) by comparing them with the points of the circle; that which is asserted of every point of the circle in the first can be asserted of the point B, because from the second B is one of these points. Again, in the former argument, whatever is asserted of every thing useful is true of geometry, because geometry is useful.

The common term of the two propositions is called the *middle term*, while the *predicate* and *subject* of the conclusion are called the *major* and *minor* terms, respectively. The two first assertions are called the *major* and *minor premisses*, and the last the *conclusion*. Suppose now the two premisses and conclusion of the syllogism just quoted to be varied in every possible way from affirmative to negative, from universal to particular, and *vice versa*, where the number of changes will be $4 \times 4 \times 4$, or 64 (called moods); since each proposition may receive four different forms, and each form of one may be compounded with any of the other two. And these may be still further varied, if instead of the middle term being the subject of the first, and the predicate of the second, this order be reversed, or if the middle term be the subject of both, or the predicate of both, which will give four different figures, as they are called, to each of the sixty-four moods above mentioned. But of these very few are correct deductions, and without entering into every case we will state some general rules, being the methods which common reason would take to ascertain the truth or falsehood of any one of them, collected and generalised.*

*Whately's *Logic*, page 76, third edition. A work which should be read by all mathematical students. [Whately's *Logic* is procurable in modern editions, many of which were, until recently, widely read in our academies and colleges. The following works in which the same material is presented in a shape more comforming to modern methods may

I. The middle term must be the same in both premisses, by what has just been observed; since in the comparison of two things with one and the same third thing, in order to ascertain their connexion or discrepancy, consists the whole of reasoning. Thus, the deduction without further process of the equation $a^2 + b^2 = c^2$ from the proposition, which proves that the sum of the *squares* described on the sides of a right-angled triangle is equal to the square on its hypothenuse, a, b, and c being the number of linear units in the sides and the hypothenuse, is incorrect, since syllogistically stated the argument would stand thus:

$$\left.\begin{array}{l}\text{The sum of the } \textit{squares} \text{ of the} \\ \quad \text{lines } a \text{ and } b \\ \qquad \text{and} \\ \text{the } \textit{square} \text{ of the line } c\end{array}\right\} \text{ are equal quantities,}$$

---

be mentioned: T. Fowler's *Elements of Deductive Logic*; Bain's *Logic*; Venn's *Empirical Logic and Symbolical Logic*; Keynes's *Formal Logic*; Carveth Read's *Logic, Deductive and Inductive*; Mill's *System of Logic* (a discussion rather than a presentation). Strictly contemporary logic will be found represented in the following works in English: Jevons's *Principles of Science and Studies in Deductive Logic*; Bradley's *Principles of Logic*; Sidgwick's *Process of Argument*; Bosanquet's *Logic: or, the Morphology of Knowledge*; and the same author's *Essentials of Logic*; Sigwart's *Logic*, recently translated from the German; and Ueberweg's *System of Logic and History of Logical Doctrines*.—*Ed.*]

$$\left.\begin{array}{c} a^2+b^2 \\ \text{and} \\ c^2 \end{array}\right\} \text{are} \left\{\begin{array}{l} \text{the sum of the } \textit{squares} \text{ of } a \text{ and } b, \\ \qquad\text{and} \\ \text{the } \textit{square} \text{ of } c. \end{array}\right.$$

$$\text{Therefore} \quad \left.\begin{array}{c} a^2+b^2 \\ \text{and} \\ c^2 \end{array}\right\} \text{are equal quantities.}$$

Here the term *square* in the major premiss has its geometrical, and in the minor its algebraical sense, being in the first a geometrical figure, and in the second an arithmetical operation. The term of comparison is not therefore the same in both, and the conclusion does not therefore follow from the premisses.

The same error is committed if all that can be contained under the middle term be not spoken of either in the major or minor premiss. For if each premiss mentions only a part of the middle term, these parts may be different, and the term of comparison really different in the two, though passing under the same name in both. Thus,

All the triangle is in the circle,
All the square is in the circle,

proves nothing, since the square may, consistently with these conditions, be either wholly, partly, or not at all contained in the triangle. In fact, as we have before shown, each of these assertions speaks of a part of the circle only. The following is of the same kind:

Some of the triangle is in the circle.
Some of the circle is not in the square, etc.

II. If both premisses are negative, no conclusion can be drawn. For it can evidently be no proof either of agreement or disagreement that two things both disagree with a third. Thus the following is inconclusive:

None of the circle is in the triangle.
None of the square is in the circle.

III. If both premisses are particular, no conclusion can be drawn, as will appear from every instance that can be taken, thus:

Some of the circle is in the triangle.
Some of the square is not in the circle,

proves nothing.

IV. In forming a conclusion, where a conclusion can be formed, nothing must be asserted more generally in the conclusion than in the premisses. Thus, if from the following,

All the triangle is in the circle,
All the circle is in the square,

we would draw a conclusion in which the square should be the subject, since the whole square is not mentioned in the minor premiss, but only part of it, the conclusion must be,

Part of the square is in the triangle.

V. If either of the premisses be negative, the conclusion must be negative. For as both premisses cannot be negative, there is asserted in one premiss an agreement between the term of the conclusion and the middle term, and in the other premiss a disagreement between the other term of the conclusion, and the same middle term. From these nothing can be inferred but a disagreement or negative conclusion. Thus, from

None of the circle is in the triangle,
All the circle is in the square,

can only be inferred,

Some of the square is *not* in the triangle.

VI. If either premiss be particular, the conclusion must be particular. For example, from

None of the circle is in the triangle,
Some of the circle is in the square,

we deduce,

Some of the square is not in the triangle.

If the student now applies these rules, he will find that of the sixty-four moods eleven only are admissible in any case; and in applying these eleven moods to the different figures he will also find that some of them are not admissible in every figure, and some not necessary, on account of the conclusion, though true, not being as general as from the premisses it might be. This he may do either by reasoning or by actual inspection of the figures, drawn and arranged according to the premisses. The admissible moods are nineteen in number, and are as follows, where *A* at the beginning of a proposition signifies that it is a universal affirmative, *E* a universal negative, *I* a particular affirmative, *O* a particular negative.

Figure I. The middle term is the subject of the major, and the predicate of the minor premiss.

1.*	*A*	All the	○ is in the △,
	A	All the	□ is in the ○,
∴	*A*	All the	□ is in the △.
2.	*E*	None of the	○ is in the △,
	A	All the	□ is in the ○,
∴	*E*	None of the	□ is in the △.
3.	*A*	All the	○ is in the △,
	I	Some of the	□ is in the ○,
∴	*I*	Some of the	□ is in the △.
4.	*E*	None of the	○ is in the △,
	I	Some of the	□ is in the ○,
∴	*O*	Some of the	□ is not in △.

Figure II. The middle term is the predicate of both pre-

*This, and 3, are the most simple of all the combinations, and the most frequently used, especially in geometry.

misses.

1.	E	None of the △ is in the ◯,
	A	All the □ is in the ◯,
∴	E	None of the □ is in the △.
2.	A	All the △ is in the ◯,
	E	None of the □ is in the ◯,
∴	E	None of the □ is in the △.
3.	E	None of the △ is in the ◯,
	I	Some of the □ is in the ◯,
∴	O	Some of the □ is not in △.
4.	A	All the △ is in the ◯,
	O	Some of the □ is not in ◯,
∴	O	Some of the □ is not in △.

Figure III. The middle term is the subject of both premisses.

1.	A	All the ◯ is in the △,
	A	All the ◯ is in the □,
∴	I	Some of the □ is in the △.
2.	I	Some of the ◯ is in the △,
	A	All the ◯ is in the □,
∴	I	Some of the □ is in the △.
3.	A	All the ◯ is in the △,
	I	Some of the ◯ is in the □,
∴	I	Some of the □ is in the △.

4. *E* None of the ◯ is in the △,
A All the ◯ is in the □,
∴ *O* Some of the □ is not in △.

5. *O* Some of the ◯ is not in △,
A All the ◯ is in the □,
∴ *O* Some of the □ is not in △.

6. *E* None of the ◯ is in the △,
I Some of the ◯ is in the □,
∴ *O* Some of the □ is not in △.

Figure IV. The middle term is the predicate of the major, and the subject of the minor premiss.

1. *A* All the △ is in the ◯,
A All the ◯ is in the □,
∴ *I* Some of the □ is in the △.

2. *A* All the △ is in the ◯,
E None of the ◯ is in the □,
∴ *E* None of the □ is in the △.

3. *I* Some of the △ is in the ◯,
A All the ◯ is in the □,
∴ *I* Some of the □ is in the △.

4. *E* None of the △ is in the ◯,
A All the ◯ is in the □,
∴ *O* Some of the □ is not in △.

5. *E* None of the △ is in the ◯,
I Some of the ◯ is in the □,
∴ *O* Some of the □ is not in △.

We may observe that it is sometimes possible to condense two or more syllogisms into one argument, thus:

Every A is B (1),
Every B is C (2),
Every C is D (3),
Every D is E (4),
Therefore Every A is E (5),

is equivalent to three distinct syllogisms of the form Fig. 1.; these syllogisms at length being (1), (2), a; a, (3), b; b, (4), (5).

The student, when he has well considered each of these, and satisfied himself, first by the rules, and afterwards by inspection, that each of them is legitimate; and also that all other moods, not contained in the above, are not allowable, or at least do not give the most general conclusion, should form for himself examples of each case, for instance of Fig. III, 3:

> The axioms constitute part of the basis of geometry.
>
> Some of the axioms are grounded on the evidence of the senses.
>
> ∴ Some evidence derived from the senses is part of the basis of geometry.

He should also exercise himself in the first principles of reasoning by reducing arguments as found in books to the syllogistic form. Any controversial or argumentative work will furnish him with a sufficient number of instances.

Inductive reasoning is that in which a universal proposition is proved by proving separately every one of its particular cases. As where, for example, a figure, $ABCD$, is proved to be a rectangle by proving each of its angles separately to be a right angle, or proving all the premisses of the following, from which the conclusion follows necessarily:

> The angles at A, B, C, and D are all the angles of the figure $ABCD$.
>
> A is a right angle,
> B is a right angle,
> C is a right angle,
> D is a right angle,
>
> Therefore all the angles of the figure $ABCD$ are right angles.

This may be considered as one syllogism of which the minor premiss is,

> A, B, C, and D are right angles,

where each part is to be separately proved.

Reasoning *à fortiori*, is that contained in Fig. I. 1. in a different form, thus: A is greater than B, B is greater than C; *à fortiori* A is greater than C; which may be also stated as follows:

The whole of B is contained in A,

The whole of C is contained in B,

Therefore C is contained in A.

The premisses of the second do not necessarily imply as much as those of the first; the complete reduction we leave to the student.

The elements of geometry present a collection of such reasonings as we have just described, though in a more condensed form. It is true that, for the convenience of the learner, it is broken up into distinct propositions, as a journey is divided into stages; but nevertheless, from the very commencement, there is nothing which is not of the nature just described. We present the following as a specimen of a geometrical proposition reduced nearly to a syllogistic form. To avoid multiplying petty syllogisms, we have omitted some few which the student can easily supply.

Hypothesis.—ABC is a right-angled triangle the right angle being at A.

Consequence.—The squares on AB and AC are together equal to the square on BC.

Construction: Upon BC and BA describe squares, produce DB to meet EF, produced, if necessary, in G, and through A draw HAK parallel to BD.

Demonstration.

I. Conterminous sides of a square are at right angles to one another. (Definition.)

EB and BA are conterminous sides of a square. (Construction.)

$\therefore$ EB and BA are at right angles.

II. A similar syllogism to prove that DB and BC are at right angles, and another to prove that GB and BC are at right angles.

III. Two right lines drawn perpendicular to two other right lines make the same angle as those others (already proved); EB and BG and AB and BC are two right lines, etc., (I. II.).

$\therefore$ The angle EBG is equal to ABC.

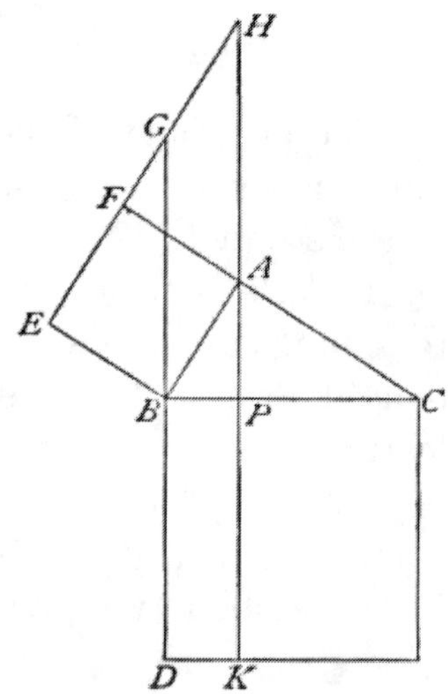

Fig. 5.

IV. All sides of a square are equal. (Definition.)

AB and BE are sides of a square. (Construction.)

$\therefore$ AB and BE are equal.

V. All right angles are equal. (Already proved.)

BEG and BAC are right angles. (Hypothesis and construction.)

$\therefore$ BEG and BAC are equal angles.

VI. Two triangles having two angles of one equal to two angles of the other, and the interjacent sides equal, are equal in all respects. (Proved.)

BEG and BAC are two triangles having BEG and EBG respectively equal to BAC and ABC and the sides EB and BA equal. (III. IV. V.)

$\therefore$ The triangles BEG, BAC are equal in all respects.

VII. BG is equal to BC. (VI.)

BC is equal to BD. (Proved as IV.)

$\therefore$ BG is equal to BD.

VIII. A four-sided figure whose opposite sides are parallel is a parallelogram. (Definition.) $BGHA$ and $BPKD$ are four-sided figures, etc. (Construction.)

$\therefore$ $BGHA$ and $BPKD$ are parallelograms.

IX. Parallelograms upon the same base and between the same parallels are equal. (Proved.) $EBAF$ and $BGHA$, are parallelograms, etc. (Construction.)

$\therefore$ $EBAF$ and $BGHA$ are equal.

X. Parallelograms on equal bases and between the same parallels, are equal. (Proved.)

$BGHA$ and $BDKP$ are parallelograms, etc. (Construction.)

$\therefore$ $BGHA$ and $BDKP$ are equal.

XI. $EBAF$ is equal to $BGHA$. (IX.)

$BGHA$ is equal to $BDKP$. (X.)

$\therefore$ $EBAF$ (that is the square on AB) is equal to $BDKP$.

XII. A similar argument from the commencement to prove that the square on AC is equal to the rectangle CPK.

XIII. The rectangles BK and CK are together equal to the square on BC. (Self-evident from the construction.)

The squares on BA and AC are together equal to the rectangles BK and CK. (Self-evident from XI. and XII.)

$\therefore$ The squares on BA and AC are together equal to the square on BC.

Such is an outline of the process, every step of which the student must pass through before he has understood the demonstration. Many of these steps are not contained in the book, because the most ordinary intelligence is sufficient to suggest them, but the least is as necessary to the process as the greatest. Instead of writing the propositions at this length, the student is recommended to adopt the plan which we now lay before him.

The explanation of this is as follows: the whole proposition is divided into distinct assertions, which are placed in separate consecutive paragraphs, which paragraphs are numbered in the first column on the left; in the second column on the left we state the reasons for each paragraph, either by referring to the preceding paragraphs from which they follow, or the preceding propositions in which they have been proved. In the latter case a letter is placed in the column, and at the end, the enunciation of the proposition there used is written opposite to the letter. By this method, the proposition is much shortened, its more prominent parts are brought immediately under notice, and the beginner, if he recollect the preceding propositions perfectly well, is not troubled by the repetition of prolix enunciations, while in the contrary

Hyp.	1		ABC is a triangle, right-angled at A.
Constr.	2	a	On BA describe a square $BAFE$.
	3	a	On BC describe a square.
	4		Produce BD to meet EF, produced if necessary, in G.
	5	b	Through A draw HAK parallel to BD.
Demonst.	6	2, Def.	EBA is a right angle.
	7	3	GBC is a right angle.
	8	$6, 7, c$	$\angle EBG$ is equal to $\angle ABC$.
	9	$2, 1, d$	$\angle BEG$ is equal to $\angle BAC$.
	10	2	EB is equal to AB.
	11	$8, 9, 10, e$	The triangles BEG and ABC are equal.
	12	11, 3	BG is equal to BD.
	13	5, 2, Def.	$AHGB$ is a parallelogram.
	14	5, 3, Def.	$BPDK$ is a parallelogram.
	15	$13, 2, f$	$AHGB$ and $ABEF$ are equal.
	16	$13, 14, g$	$AHGB$ and $BPDK$ are equal.
	17	15, 16	$BPDK$ and the square on AB are equal.
	18	By similar reasoning	CPK and the square on CA are equal.
	19	17, 18	The square on BC is equal to the squares on BA and AC.

a, b	Here refer to the necessary problems.
c	If two lines be drawn at right angles to two others, the angles made by the first and second pair are equal.
d	All right angles are equal.
e	Two triangles which have two angles of one equal to two angles of the other, and the interjacent sides equal, are equal in all respects.
f, g	Parallelograms on the same or equal bases, and between the same parallels, are equal.

case he has them at hand for reference.

In all that has been said, we have taken instances only of direct reasoning, that is, where the required result is immediately obtained without any reference to what might have happened if the result to be proved had not been true. But there are many propositions in which the only possible result is one of two things which cannot be true at the same time, and it is more easy to show that one is *not* the truth, than that the other *is*. This is called indirect reasoning; not that it is less satisfactory than the first species, but because, as its name imports, the method does not appear so direct and natural. There are two propositions of which it is required to show that whenever the first is true the second is true; that is, the first being the hypothesis the second is a necessary conclusion from it, whence the hypothesis in question, and anything contradictory to, or inconsistent with, the conclusion cannot exist together. In indirect reasoning, we suppose that, the original hypothesis existing and being true, something inconsistent with or contradictory to the conclusion is true also. If from combining the consequences of these two suppositions, something evidently erroneous or absurd is deduced, it is plain that there is something wrong in the assumptions. Now care is taken that the only doubtful point shall be the one just alluded to, namely, the supposition that one proposition and the contradictory of the other are true together. This then is incorrect, that is, the first proposition cannot exist with anything contradictory to the second, or the second must exist wherever the first exists, since if any proposition be not true its contradictory must be true, and *vice versa.* This is rather embarrassing to the beginner, who finds that he is required to admit, for argument's sake, a proposition which the argument itself goes to destroy. But the difficulty would be materially lessened, if instead of assuming the contradictory of the second proposition positively, it were hypothetically stated, and the consequences of it asserted with the verb "would be," instead of "is." For example: suppose it to be known that if A is B, then C must be D, and it is required to show indirectly that when C is not D, A is not B. This put into the form in which such a proposition would appear in most elementary works, is as follows.

It being granted that if A is B, C is D, it is required to show that when C is not D, A is not B. If possible, let C be not D, and let A be B. Then by what is granted, since

A is B, C is D; but by hypothesis C is not D, therefore both C *is* D and *is not* D, which is absurd; that is, it is absurd to suppose that C *is not* D and A *is* B, consequently when C is not D, A is not B. The following, which is exactly the same thing, is plainer in its language. Let C be not D. Then if A *were* B, C *would be* D by the proposition granted. But by hypothesis C is not D, etc. This sort of indirect reasoning frequently goes by the name of *reductio ad absurdum*.

In all that has gone before we may perceive that the validity of an argument depends upon two distinct considerations,—(1) the truth of the relations assumed, or represented to have been proved before; (2) the manner in which these facts are combined so as to produce new relations; in which last the *reasoning* properly consists. If either of these be incorrect in any single point, the result is certainly false; if both be incorrect, or if one or both be incorrect in more points than one, the result, though not at all to be depended on, is not certainly false, since it may happen and has happened, that of two false reasonings or facts, or the two combined, one has reversed the effect of the other and the whole result has been true; but this could only have been ascertained after the correction of the erroneous fact or reasoning. The same thing holds good in every species of reasoning, and it must be observed, that however different geometrical argument may be in form from that which we employ daily, it is not different in reality. We are accustomed to talk of mathematical *reasoning* as above all other, in point of accuracy and soundness. This, if by the term *reasoning* we mean the comparing together of different ideas and producing other ideas from the comparison, is not correct, for in this view mathematical reasonings and all other reasonings correspond exactly. For the real difference between mathematics and other studies in this respect we refer the student to the first chapter of this treatise.

In what then, may it be asked, does the real advantage of mathematical study consist? We repeat again, in the actual certainty which we possess of the truth of the facts on which the whole is based, and the possibility of verifying every result by actual measurement, and not in any superiority which the method of reasoning possesses, since there is but one method of reasoning. To pursue the illustration with which we opened this work (page the first), suppose this point to be raised, was the slaughter of Cæsar justifiable or not? The actors in that deed justified themselves by saying, that a tyrant and usurper, who meditated the destruction of his country's liberty, made it the duty of every citizen to put him to death, and that Cæsar was a tyrant and usurper, etc. Their *reasoning* was perfectly correct, though proceeding on premisses then extensively, and now universally, denied. The first premiss, though correctly used in this reasoning, is now asserted to be false, on the ground that it is the duty of every citizen to do nothing which would, were the practice universal, militate against the general happiness; that were each individual to act upon his own judgment, instead of leaving offenders to the law, the result would be anarchy and complete destruction of civilisation, etc. Now in these reasonings and all others, with the exception of those which occur in mathematics, it must be observed that there are no premisses so certain, as never to have been denied, no first principles to which the same degree of evidence is attached as to the following, that "no two straight lines can enclose a space." In mathematics, therefore, we reason on certainties, on notions to which the name of innate can be applied, if it can be applied to any whatever. Some, on observing that we dignify such simple consequences by the name of reasoning, may be loth to think that this is the process to which they used to attach such ideas of difficulty. There may, perhaps, be many who imagine that reasoning is for the mathematician, the logician, etc., and who, like the Bourgeois Gentilhomme, may be surprised on being told, that, well or ill, they have been reasoning all their lives. And yet such is the fact; the commonest actions of our lives are directed by processes exactly identical with those which enable us to pass from one proposition of geometry to another. A porter, for example, who being directed to carry a parcel from the city to a street which he has never heard of, and who on inquiry, finding it is in the Borough, concludes that he must cross the water to get at it, has performed an act of reasoning, differing nothing in kind from those by a series of which, did he know the previous propositions, he might be convinced that the square of the hypothenuse of a right-angled triangle is equal to the sum of the squares of the sides.

CHAPTER XV.

ON AXIOMS.

GEOMETRY, then, is the application of strict logic to those properties of space and figure which are self-evident, and which therefore cannot be disputed. But the rigor of this science is carried one step further; for no property, however evident it may be, is allowed to pass without demonstration, if that can be given. The question is therefore to demonstrate all geometrical truths with the smallest possible number of assumptions. These assumptions are called axioms, and for an axiom it is requisite: (1) that it should be self-evident; (2) that it should be incapable of being proved from the other axioms. In fulfilling these conditions, the number of axioms which are really geometrical, that is, which have not equal reference to Arithmetic, is reduced to two, viz., two straight lines cannot enclose a space, and through a given point not more than one parallel can be drawn to a given straight line. The first of these has never been considered as open to any objection; it has always passed as perfectly self-evident.* It is on this account made the proposition on which

*But see J. B. Stallo, *Concepts and Theories of Modern Physics*, New York, 1884, p. 242, p. 208 et seq., and p. 248 et seq. For popular philosophical discussions of the subject of Axioms generally, in the light of modern psychology and pangeometry, the reader may consult the following works: Helmholtz's "Origin and Meaning of Geometrical Axioms," *Mind*, Vol. III., p. 215, and the article in the same author's *Popular Lectures on Scientific Subjects*, Second Series, London, 1881, are grounded all reasonings relative to the straight line, since the definition of a straight line is too vague to afford any information. But the second, viz., that through a given point not more than one parallel can be drawn to a given straight line, has always been considered as an assumption not self-

---

pp. 27–71; W. K. Clifford's *Lectures and Essays*, Vol. I., p. 297, p. 317; Duhamel, *Des Méthodes dans les Sciences de Raisonnement*, Part 2; and the articles "Axiom" and "Measurement" in the *Encyclopedia Britannica*, Vol. XV. See also Riemann's Essay on the *Hypotheses Which Lie at the Basis of Geometry*, a translation of which is published in Clifford's *Works*, pp. 55–69. For part of the enormous technical literature of this subject cf. Halsted's *Bibliography of Hyper-Space and Non-Euclidean Geometry*, *American Journal of Mathematics*, Vol. I., pp. 261 et seq., and Vol. II., pp. 65 et seq. Much, however, has been written subsequently to the date of the last-mentioned compilation, and translations of Lobachévski and Bolyai, for instance, may be had in the *Neomonic Series* of Dr. G. B. Halsted (Austin, Texas). A full history of the theory of parallels till recent times is given in Paul Stäckel's *Theorie der Parallellinien von Euklid bis auf Gauss* (Leipsic, 1895). Of interest are the essays of Prof. J. Delbœuf on *The Old and the New Geometries* (*Revue Philosophique*, 1893–1895), and those of Professor Poincaré and of other controversialists in the recent volumes of the *Revue de Métaphysique et de Morale*, where valuable bibliographical references will be found to literature not mentioned in this note. See also P. Tannery in the recent volumes of the *Revue générale* and the *Revue philosophique*, Poincaré in *The Monist* for October, 1898, and B. A. W. Russell's *Foundations of Geometry* (Cambridge, 1897). In Grassmann's *Ausdehnungslehre* (1844), "assumptions" and "axioms" are replaced by purely formal (logical) "predications," which presuppose merely the consistency of mental operations. (See *The Open Court*, Vol. II. p. 1464, Grassmann, "A Flaw in the Foundation of Geometry," and Hyde's *Directional Calculus*, Ginn & Co., Boston). Dr. Paul Carus in his *Primer* evident in itself, and has therefore been called the defect and disgrace of geometry. We proceed to place it on what we conceive to be the proper footing.

By taking for granted the arithmetical axioms only, with the first of those just alluded to, the following propositions may be strictly shown.

I. One perpendicular, and only one, can be let fall from any point A to a given line CD. Let this be AB.

II. If equal distances BC and BD be taken on both sides of B, AC and AD are equal, as also the angles BAC and BAD.

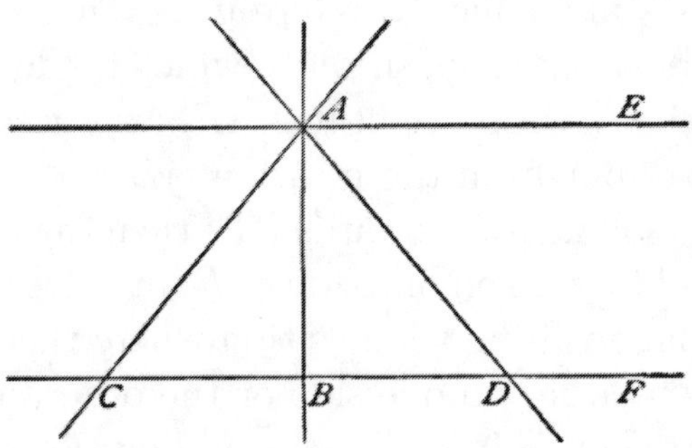

Fig. 6.

III. Whatever may be the length of BC and BD, the

of Philosophy (Chicago), p. 51 et seq., has treated the subject of Axioms at length, from a similar point of view. On the psychological side, consult Mach's *Analysis of the Sensations* (Chicago, 1897), and the bibliographical references and related discussions in such works as James's *Psychology* and Jodl's *Psychology* (Stuttgart, 1896).—*Ed.*

angles BAC and BAD are each less than a right angle.

IV. Through A a line may be drawn parallel to CD (that is, by definition, never meeting CD, though the two be ever so far produced), by drawing any line AD and making the angle DAE equal to the angle ADB, which it is before shown how to do.

From proposition IV. we should at first see no reason against there being as many parallels to CD, to be drawn through A, as there are different ways of taking AD, since the direction for drawing a parallel to CD is, "take *any line* AD cutting CD and make the angle DAE equal to ADB." But this our senses immediately assure us is impossible.

It appears also a proposition to which no degree of doubt can attach, that if the straight line AB, produced indefinitely both ways, set out from the position AB and revolve round the point A, moving first towards AE; then the point of intersection D will first be on one side of B and afterwards on the other, and there will be one position where there is no point of intersection either on one side or the other, and *one such position only.* This is in reality the assumption of Euclid; for having proved that AE and BF are parallel when the angles BDA and DAE are equal, or, which is the same thing, when EAD and ADF are together equal to two right angles, he further assumes that they will be parallel in no other case, that is, that they will meet when the angles EAD and ADF are together greater or less than two right angles; which is really only assuming that the parallel which he has found is the only one which can be drawn. The remaining part of his axiom, namely, that the lines AE and DF, if they meet at all, will meet upon that side of DA on which the angles are less than two right angles, is not an assumption but a consequence of his proposition which shows that any two angles of a triangle are together less than two right angles, and which is established before any mention is made of parallels. It has been found by the experience of two thousand years that some assumption of this sort is indispensable. Every species of effort has been made to avoid or elude the difficulty, but hitherto without success, as some assumption has always been involved, at least equal, and in most cases superior, in difficulty to the one already made by Euclid. For example, it has been proposed to define parallel lines as those which are equidistant from one another at every point. In this case, before the name *parallel* can be allowed to belong to any thing, it must be proved that there are lines such that a perpendicular to one is always perpendicular to the other, and that the parts of these perpendiculars intercepted between the two are always equal. A proof of this has never been given without the previous assumption of something equivalent to the axiom of Euclid. Of this last, indeed, a proof has been given, but involving considerations not usually admitted into geometry, though it is more than probable that had the same come down to us, sanctioned by the name of Euclid, it would have been received without difficulty. The Greek geometer confines his notion of equal magnitudes to those which have boundaries. Suppose this notion of equality extended to all such spaces as can be made to coincide entirely in all their extent, what ever that extent may be; for example, the unbounded spaces contained between two equal angles whose sides are produced without end, which by the definition of equal angles might be made to coincide entirely by laying the sides of one angle upon those of the other. In the same sense we may say, that, one angle being double another, the space contained by the sides of the first is double that contained by the sides of the second, and so on. Now suppose two lines Oa and Ob, making any angle with one

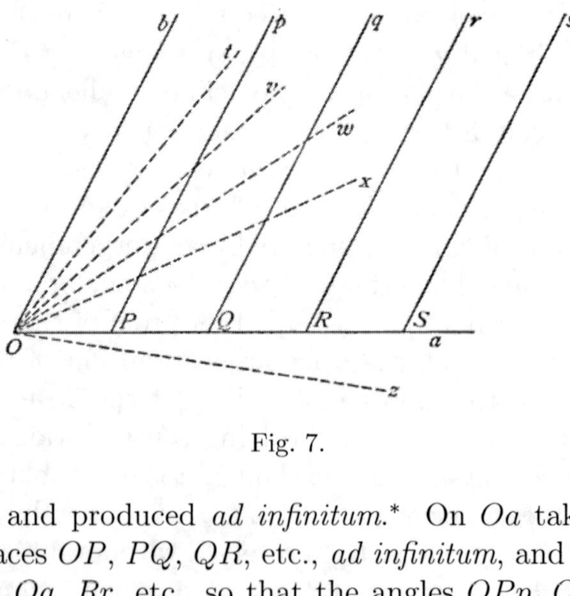

Fig. 7.

another, and produced *ad infinitum.** On Oa take off the equal spaces OP, PQ, QR, etc., *ad infinitum*, and draw the lines Pp, Qq, Rr, etc., so that the angles OPp, OQq, etc.,

*Every line in this figure must be produced *ad infinitum*, from that extremity at which the small letter is placed.

shall be equal to one another, each being such as with bOP will make two right angles. Then Ob, Pp, Qq, etc., are parallel to one another, and the infinite spaces $bOPp$, $pPQq$, $qQRr$, etc., can be made to coincide, and are equal. Also no finite number whatever of these spaces will fill up the infinite space bOa, since OP, PQ, etc., may be contained *ad infinitum* upon the line Oa. Let there be any line Ot, such that the angles tOP and pPO are together less than two right angles, that is, less than bOP and pPO; whence tOP is less than bOP and tO falls between bO and aO. Take the angles tOv, vOw, wOx, each equal to bOt, and continue this until the last line Oz falls beneath Oa, so that the angle bOz is greater than bOa. That this is possible needs no proof, since it is manifest that any angle being continually added to itself the sum will in time exceed any other given angle; again, the infinite spaces bOt, tOv, etc., are all equal. Now on comparing the spaces bOt and $bOPp$, we see that a certain number of the first is more than equal to the space bOa, while no number whatever of the second is so great. We conclude, therefore, that the space bOt is greater than $bOPp$, which cannot be unless the line Ot cuts Pp at last; for if Ot did never cut Pp, the space bOt would evidently be less than $bOPp$, as the first would then fall entirely within the second. Therefore two lines which make with a third angles together less than two right angles will meet if sufficiently produced. [See Note on page 241.]

This demonstration involves the consideration of a new species of magnitude, namely, the whole space contained by the sides of an angle produced without limit. This space is unbounded, and is greater than any number whatever of finite spaces, of square feet, for example. No comparison, therefore, as to magnitude can be instituted between it and any finite space whatever, but that affords no reason against comparing this magnitude with others of the same kind.

Any thing may become the subject of mathematical reasoning, which can be increased or diminished by other things of the same kind; this is, in fact, the definition given of the term magnitude; and geometrical reasoning, in all other cases at least, can be applied as soon as a criterion of equality is discovered. Thus the angle, to beginners, is a perfectly new species of magnitude, and one of whose measure they have no conception whatever; they see, however, that it is capable of increase or diminution, and also that two of the kind can be equal, and how to discover whether this is so or not, and nothing more is necessary for them. All that can be said of the introduction of the angle in geometry holds with some, (to us it appears an equal force,) with regard to these unlimited spaces; the two are very closely connected, so much so, that the term angle might even be defined as "the unlimited space contained by two right lines," without alteration in the truth of any theorem in which the word *angle* is found. But this is a point which cannot be made very clear to the beginner.

The real difficulties of geometry begin with the theory of proportion, to which we now proceed. The points of discussion which we have hitherto raised, are not such as to embarrass the elementary student, however much they may perplex the metaphysical inquirer into first principles. The theory to which we are coming abounds in difficulties of both classes.

[NOTE TO PAGE 239.—The demonstration given on pp. 237–239 is now regarded as fallacious by mathematicians; the considerations that apply to finite aggregates not being transferable to infinite aggregates,—for example, it is not true for infinite aggregates that the part is always less than the whole. Even Plato is cited for the assertion that equality is only to be predicated of finite magnitudes. See the modern works on the Theory of the Infinite. The demonstration in question is not De Morgan's, but M. Bertrand's.—*Ed.*]

CHAPTER XVI.

ON PROPORTION.

In the first elements of geometry, two lines, or two surfaces, are mentioned in no other relation to one another than that of equality or non-equality. Nothing but the simple fact is announced that one magnitude is equal to, greater than, or less than another, except occasionally when the sum of two equal magnitudes is said to be double one of them. Thus in proving that two sides of a triangle are together greater than the third, the fact that they are *greater* is the essence of the proposition; no measure is given of the excess, nor does anything follow from the theorem as to whether it is, or may be, small or great. We now come to the doctrine of proportion in which geometrical magnitude is considered in a new light. The subject has some difficulties, which have been materially augmented by the almost universal use, in this country at least,* of the theory laid down in the fifth book of Euclid.† Considered as a complete conquest over a great and acknowledged difficulty of principle, this book of Euclid well deserves the immortality of which its existence, at the present moment, is the guarantee; nay, had the speculations of the mathematician been wholly confined to geometrical magnitude, it might be a question whether any other notions would be necessary. But when we come

*In England.

†See Todhunter's *Euclid* (Macmillan, London).—*Ed.*

to apply arithmetic to geometry, it is necessary to examine well the primary connexion between the two; and here difficulties arise, not in comprehending that connexion so much as in joining the two sciences by a chain of demonstration as strong as that by which the propositions of geometry are bound together, and as little open to cavil and disputation.

The student is aware that before pronouncing upon the connexion of two lines with one another, it is necessary to *measure* them, that is, to refer them to some third line, and to observe what number of times the third is contained in the other two. Whether the two first are equal or not is readily ascertained by the use of the compasses, on principles laid down with the utmost strictness in Euclid and other elementary works. But this step is not sufficient; to say that two lines are not equal, determines nothing. There are an infinite number of ways in which one line may be greater or less than a given line, though there is only one in which the other can be equal to the given one. We proceed to show how, from the common notion of measuring a line, the more strict geometrical method is derived.

To measure the line AB, apply to it another line (the edge of a ruler), which is divided into equal parts (as inches), each of which parts is again subdivided into ten equal parts, as in the figure. This division is made to take place in practice until the last subdivision gives a part so small that anything less may be neglected as inconsiderable. Thus a carpenter's rule is divided into tenths or eighths of inches only, while in the tube of a barometer a process must be employed which will mark a much less difference. In talking of accurate measurement, therefore, anywhere but in geometry, or algebra, we only mean accurate as far as the senses are concerned, and as far as is necessary for the object in view. The ruler in the figure shows that the line AB contains more than two and less than three inches; and closer inspection shows that the excess above two inches is more than sixth-tenths of an inch, and less than seven. Here, in practice, the process stops; for, as the subdivision of the ruler was carried only to tenths of inches, because a tenth of an inch is a quantity which may be neglected in ordinary cases, we may call the line two inches and six-tenths, by doing which the error committed is less than one-tenth of an inch. In this way lines may be compared together with a common degree of correctness; but this is not enough for the geometer. His notions of accuracy are not confined to tenths or hundredths, or hundred-millionth parts of any line, however small it may be at first. The reason is obvious; for although to suit the eye of the generality of readers, figures are drawn in which the least line is usually more than an inch, yet his theorems are asserted to remain true, even though the dimensions of the figure are so far diminished as to make the whole imperceptible in the strongest microscope. Many theorems are obvious upon looking at a moderately-sized figure; but the reasoning must be such as to convince the mind of their truth when,

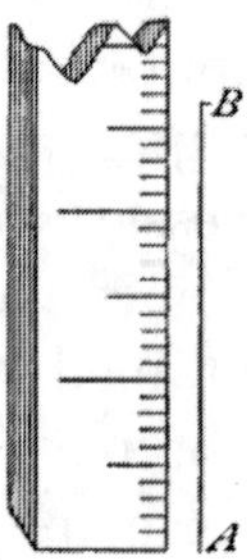

Fig. 8.

from excessive increase or diminution of the scale, the figures themselves have past the boundary even of imagination. The next step in the process of measurement is as follows, and will lead us to the great and peculiar difficulty of the subject.

The inch, the foot, and the other lengths by which we compare lines with one another, are perfectly arbitrary. There is no reason for their being what they are, unless we adopt the commonly received notion that our inch is derived from our Saxon ancestors, who observed that a barley corn is always of the same length, or nearly so, and placed three of them together as a common standard of measure, which they called an inch. Any line whatever may be chosen as the standard of measure, and it is evident that when two or more lines are under consideration, exact comparisons of their lengths can only be obtained from a line which is contained an exact number of times in them all. For even exact fractional measures are reduced to the same denominator, in order to compare their magnitudes. Thus, two lines which contain $\frac{2}{11}$ and $\frac{3}{7}$ of a foot, are better compared by observing that $\frac{2}{11}$ and $\frac{3}{7}$ being $\frac{14}{77}$ and $\frac{33}{77}$, the given lines contain one 77th part of a foot 14 and 33 times respectively. Any line which is contained an exact number of times in another is called in geometry a measure of it, and a common measure of two or more lines is that which is contained an exact number of times in each.

Again, a line which is measured by another is called a multiple of it, as in arithmetic.

The same definition, *mutatis mutandis*, applies to surfaces, solids, and all other magnitudes; and though in our succeeding remarks we use lines as an illustration, it must be recollected that the reasoning applies equally to every magnitude which can be made the subject of calculation.

In order that two quantities may admit of comparison as to magnitude, they must be of the same sort; if one is a line, the other must be a line also. Suppose two lines A and B each of which is measured by the line C; the first containing it five times and the second six. These lines A and B, which contain the same line C five and six times respectively, are said to have to one another the ratio of five to six, or to be in the proportion of five to six. If then we denote the first by A,* and the second by B, and the common measure by C,

*The student must distinctly understand that the common meaning of algebraical terms is departed from in this chapter, wherever the letters are large instead of small. For example, A, instead of meaning the number of *units* of some sort or other contained in the line A, stands for *the line A itself*, and mA (the small letters throughout meaning *whole numbers*) stands for the line made by taking A, m times. Thus such expressions as $mA+B$, $mA-nB$, etc., are the only ones admissible. AB, $\frac{A}{B}$, A^2, etc., are unmeaning, while $\frac{A}{m}$ is the line which is contained m times in A, or the mth part of A. The capital letters throughout stand for concrete quantities, not for their representations in abstract numbers.

we have

$$\begin{aligned} A &= 5C, \quad \text{or} \quad 6A = 30C, \\ B &= 6C, \quad \text{or} \quad 5B = 30C, \\ \text{whence} \quad 6A &= 5B, \quad \text{or} \quad 6A - \ \ 5B = 0. \end{aligned}$$

Generally, when $mA - nB = 0$, the lines, or whatever they are, represented by A and B, are said to be in the proportion of n to m, or to have the ratio of n to m.

Let there be two other magnitudes P and Q, of the same kind with one another, either differing from the first in kind or not, (thus A and B may be lines, and P and Q surfaces, etc.,) and let them contain a common measure R, just as A and B contain C, viz.: Let P contain R five times, and let Q contain R six times, we have by the same reasoning

$$6P - 5Q = 0,$$

and P and Q, being also in the ratio of five to six, as well as A and B, are said to be proportional to A and B, which is denoted thus

$$A : B :: P : Q,$$

by which *at present* all we mean is this, that there are some two whole numbers m and n such that, at the same time

$$\begin{aligned} mA - nB &= 0, \\ mP - nQ &= 0. \end{aligned}$$

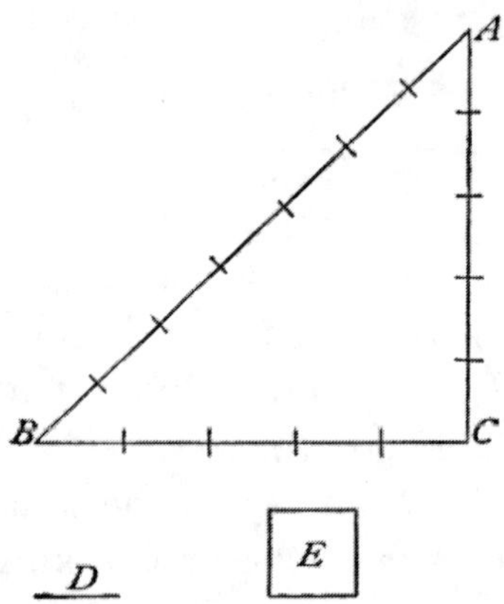

Fig. 9.

Nothing more than this would be necessary for the formation of a complete theory of proportion, if the common measure, which we have supposed to exist in the definition, did always really exist. We have, however, no right to assume that two lines A and B, whatever may be their lengths, both contain some other line an exact number of times. We can, moreover, produce a direct instance in which two lines have no common measure what ever, in the following manner.

Let ABC be an isosceles right-angled triangle, the side BC and the hypothenuse have no common measure whatever. If possible let D be a common measure of BC and AB; let BC contain D, n times, and let AB contain D, m times. Let E be the square described on D. Then since AB contains D, m times, the square described on AB contains, $m \times m$ or m^2 times. Similarly the square described on BC contains E $n \times n$ or n^2 times. But, because AB is an isosceles right-angled triangle, the square on AB is double that on BC, whence $m \times m = 2(n \times n)$ or $m^2 = 2n^2$. To prove the impossibility of this equation (when m and n are whole numbers), observe that m^2 must be an even number, since it is twice the number n^2. But $m \times m$ cannot be an even number unless m is an even number, since an odd number multiplied by itself produces an odd number.* Let m (which has been shown to be even) be double m' or $m = 2m'$. Then $2m' \times 2m' = 2n^2$ or $4m'^2 = 2n^2$ or $n^2 = 2m'^2$. By repeating the same reasoning we show that n is even. Let it be $2n'$. Then $2n' \times 2n' = 2m'^2$ or $m'^2 = 2n'^2$. By the same reasoning m' and n' are both even, and so on *ad infinitum.* This reasoning shows that the whole numbers which satisfy the equation $n^2 = 2m^2$ (if such there be) are divisible by 2 without remainder, *ad infinitum.* The absurdity of such a supposition is manifest: there are then no such whole numbers, and consequently no common measure to BA and BC.

Before proceeding any further, it will be necessary to establish the following proposition.

If the greater of two lines A and B be divided into m equal parts, and one of these parts be taken away; if the remainder be then divided into m equal parts, and one of them be taken away, and so on,—the remainder of the line A shall in time become less than the line B, how small soever the line B may be.

Take a line which is less than B, and call it C. It is evident that, by a continual addition of the same quantity

*Every odd number, when divided by 2, gives a remainder 1, and is therefore of the form $2p + 1$ where p is a whole number. Multiply $2p+1$ by itself, which gives $4p^2+4p+1$, or $2(2p^2+2p)+1$, which is an odd number, since, when divided by 2, it gives the quotient $2p^2 + 2p$, a whole number, and the remainder 1.

to C, this last will come in time to exceed A; and still more will it do so if the quantity added to C be increased at each step. To simplify the proof we suppose that 20 is the number of equal parts into which A and its remainders are successively divided, so that 19 out of the 20 parts remain after subtraction.

Divide C into 19 equal parts and add to C a line equal to one of these parts. Let the length of C, so increased, be C'. Divide C' into 19 equal parts and let C', increased by its 19th part, be C''. Now, since we add more and more each time to C, in forming C', C'', etc, we shall in time exceed A. Let this have been done, and let D be the line so obtained, which is greater than A. Observe now that C' contains 19, and C'', 20 of the same parts, whence C' is made by dividing C'' into 20 parts and removing one of them. The same of all the rest. Therefore we may return from D to C by dividing D into 20 parts, removing one of them, and repeating the process continually. But C is less than B by hypothesis. If then we can, by this process, reduce D below B, still more can we do so with A, which is less than D, by the same method.

This depends on the obvious truth, that if, at the end of any number of subtractions (D being taken), we have left $\frac{p}{q}D$, at the end of the same number of subtractions (A being taken), we shall have $\frac{p}{q}A$, since the method pursued in both cases is the same. But since A is less than D,

$\frac{p}{q} A$ is less than $\frac{p}{q} D$, which becomes equal to C, therefore $\frac{p}{q} A$ becomes less than C.*

We now resume the isosceles right-angled triangle. The lines BC and AB, which were there shown to have no common measure, are called *incommensurable* quantities, and to their existence the theory of proportion owes its difficulties. We can nevertheless show that A and B being incommensurable, a line can be found as near to B as we please, either greater or less, which is commensurable with A. Let D be any line taken at pleasure, and therefore as small as we please. Divide A into two equal parts, each of those parts into two equal parts, and so on. We shall thus at last find a part of A which is less than D. Let this part be E, and let it be contained m times in A. In the series E, $2E$, $3E$, etc., we shall arrive at last at two consecutive terms, pE and $(p+1)E$ of which the first is less, and the second greater than B. Neither of these differs from B by so much as E; still less by so

*Algebraically, let a be the given line, and let $\frac{1}{m}$th part of the remainder be removed at every subtraction. The first quantity taken away is $\frac{a}{m}$ and the remainder $a - \frac{a}{m}$ or $a\left(1 - \frac{1}{m}\right)$, whence the second quantity removed is $\frac{a}{m}\left(1 - \frac{1}{m}\right)$, and the remainder $\left(1 - \frac{a}{m}\right)\left(1 - \frac{1}{m}\right)$ or $a\left(1 - \frac{1}{m}\right)^2$. Similarly, the nth remainder is $a\left(1 - \frac{1}{m}\right)^n$. Now, since $1 - \frac{1}{m}$ is less than unity, its powers decrease, and a power of so great an index may be taken as to be less than any given quantity.

much as D; and both pE and $(p+1)E$ are commensurable with A, that is with mE, since E is a common measure of both. If therefore A and B are incommensurable, a third magnitude can be found, either greater or less than B, differing from B by less than a given quantity, which magnitude shall be commensurable with A.

We have seen that when A and B are incommensurable, there are no whole values of m and n, which will satisfy the equation $mA - nB = 0$; nevertheless, we can prove that values of m and n can be found which will make $mA - nB$ less than any given magnitude C, of the same kind, how small soever it may be. Suppose, that for certain values of m and n,* we find $mA - nB = E$, and let the first multiple of E, which is greater than B, be pE, so that $pE = B + E'$ where E' is less than E, for were it greater, $(p-1)E$ or

*It is necessary here to observe, that in speaking of the expression $mA - nB$ we more frequently refer to its form than to any actual value of it, derived from supposing m and n to have certain known values. When we say that $mA - nB$ can be made smaller than C, we mean that some values can be given to m and n such that $mA - nB < C$, or that *some* multiple of B subtracted from some multiple of A is less than C. The following expressions are all of the same form, viz., that of some multiple of B subtracted from some multiple of A:

$$mA - nB,$$
$$mpA - (np+1)B,$$
$$2mA - 4mB, \text{ etc., etc.}$$

$pE - E$, which is $B + (E' - E)$, would be greater than B, which is against the supposition.

The equation $mA - nB = E$ gives

$$pmA - pnB = pE = B + E',$$

whence

$$pmA - (pn + 1)B = E'.$$

Let

$$pm = m' \quad \text{and} \quad pn + 1 = n',$$

whence

$$m'A - n'B = E'.$$

We have therefore found a difference of multiples which is less than E. Let $p'E'$ be the first multiple of E' which is greater than B, where p' must be *at least as great as* p, since E being greater than E', it cannot take *more** of E than of E' to exceed B. Let

$$p'E' = B + E'',$$

then, as before,

$$m'p'A - (n'p' + 1)B = E'',$$

or

$$m''A - n''B = E'';$$

*It may require *as many*. Thus it requires as many of 7 as of 8 to exceed 33, though 7 is less than 8.

we have therefore still further diminished the difference of the multiples; and the process may be repeated any number of times; it only remains to show that the diminution may proceed to any extent.

This will appear superfluous to the beginner, who will probably imagine that a quantity diminished at every step, must, by continuing the number of steps, at last become as small as we please. Nevertheless if any number, as 10, be taken and its square root extracted, and the square root of that square root, and so on, the result will not be so small as unity, although ten million of square roots should have been extracted. Here is a case of continual diminution, in which the diminution is not *without limit.* Again, from the point D in the line AB draw DE, making an angle with AB less than half a right angle. Draw BE perpendicular to AB,

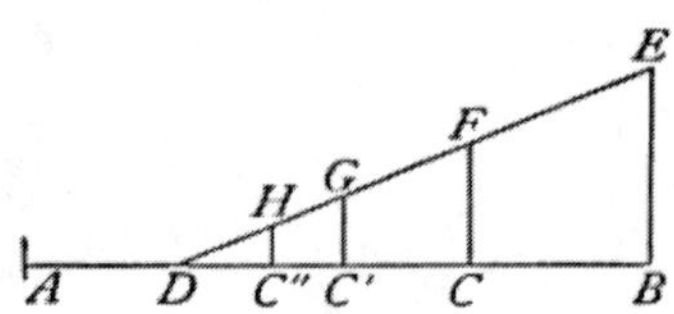

Fig. 10.

and take $BC = BE$. Draw CF perpendicular to AB, and take $CC' = CF$, and so on. The points C, C', C'', etc., will always be further from A than D is; and all the lines AC, AC', AC'', etc., though diminished at every step, will always remain greater than AD. Some such species of diminution, for anything yet proved to the contrary, may take place in $mA - nB$.

To compare the quantities E, E', etc., we have the equations

$$\begin{aligned} pE &= B + E', \\ p'E' &= B + E'', \\ p''E'' &= B + E''', \\ \text{etc.,} &\quad \text{etc.} \end{aligned}$$

The numbers p, p', p'', etc., do not diminish; the lines E, E', E'', etc., diminish at every step. If then we can show that p, p', etc., can only remain the same for a finite number of steps, and must then increase, and after the increase can only remain the same for another finite number of steps, and then must increase again, and so on, we show that the process can be continued, until one of them is as great as we please; let this be $p^{(z)}$, where z is not an exponent, but marks the number which our notation will have reached, and indicates the $(z+1)^{\text{th}}$ step of the process. Let $E^{(z)}$ be the corresponding remainder from the former step. Then, since $p^{(z)}E^{(z)}$ is the first multiple of $E^{(z)}$, which exceeds the given quantity B, if $p^{(z)}$ can be as great as we please, $E^{(z)}$ can be as small as we please. To show that $p^{(z)}$ can be as great as we please, observe, that p, p', p'', etc., must remain the same, or increase, since, as appears from their method of formation, they cannot diminish. Let them remain the same for some steps, that is, let $p = p' = p''$, etc. The equations become

$$\begin{aligned} pE &= B + E', \\ pE' &= B + E'', \\ pE'' &= B + E''', \\ \text{etc.,} &\quad \text{etc.} \end{aligned}$$

Then by subtraction,

$$\begin{aligned} E' - E'' &= p(E - E'), \\ E'' - E''' &= p(E' - E'') = pp(E - E'), \\ E''' - E'''' &= p(E'' - E''') = ppp(E - E'), \\ \text{etc.,} &\qquad\qquad \text{etc.} \end{aligned}$$

Now,

$$\begin{aligned} E - E'' &= E - E' + E' - E'' &&= (E - E')(1 + p), \\ E - E''' &= E - E' + E' - E'' + E'' - E''' &&= (E - E')(1 + p + p^2), \\ \text{etc.,} &\qquad \text{etc.,} &&\qquad \text{etc.} \end{aligned}$$

Generally,

$$\begin{aligned} E - E^{(w)} &= E - E' + E' - E'' + \cdots + E^{(w-1)} - E^{(w)} \\ &= (E - E')(1 + p + p^2 + \cdots + p^{w-1}), \end{aligned}$$

which is derived from w steps of the process. Now, if this can go on *ad infinitum*, it can go on until $1 + p + p^2 + \cdots + p^{w-1}$ is as great as we please; for, since p is not less than unity, the continual addition of its powers will, in time, give a sum exceeding any *given* number. This is absurd, from the step at which $1+p+p^2+\cdots+p^{w-1}$ becomes greater than the number of times which $E - E'$ is contained in E; for, from the above equation, $E-E'$ is contained in $E-E^{(w)}$, $1+p+p^2+\cdots+p^{w-1}$ times; and it is contradictory to suppose that $E - E'$ should be contained in $E - E^{(w)}$ more times than it is contained in E.

To take an example: suppose that B is 55 feet, and E is 54 feet; the first equation is

$$2 \times 54^f = 55^f + 53^f,$$

where $E' = 53^f$ and $E - E' = 1^f$, and is contained in E 54 times. If, then, we continue the process, 2 cannot maintain its present place through so many steps of the process as will, if the same number of terms be taken, give $1 + 2 + 2^2 + 2^3 +$ etc., greater than 54; that is, it cannot be the same for *six* steps. And we find, on actually performing the operations,

$$\begin{aligned} 2 \times 54^f &= 55 + 53^f, \\ 2 \times 53^f &= 55 + 51^f, \\ 2 \times 51^f &= 55 + 47^f, \\ 2 \times 47^f &= 55 + 39^f, \\ 2 \times 39^f &= 55 + 23^f, \\ 3 \times 23^f &= 55 + 14^f. \end{aligned}$$

We do not say that p, p', etc., *will* remain the same until $1 + p + p^2 + \dots$ would be greater than the number of times which E contains $E - E'$, but only that they cannot remain the same longer. By repetition of the same process, we can show that a further and further increase must take place, and so on until we have attained a quantity greater than any given one. And it has already been shown to be a consequence of this, that $mA - nB$ can be diminished to any extent we please. Similarly it may be shown that when A and B are incommensurable, $mA - nB$ may be brought as near as we please to any other quantity C, of the same kind as A and B, so as not to differ from C by so much as a given quantity E. For let m and n be taken, by the last case, so that $mA - nB$ may be less than E, and let $mA - nB$, in this case, be equal to E'. Let C lie between pE' and $(p + 1)E'$, neither of which can differ from C by so much as E', and therefore not by so much as E. Then since

$$mA - nB = E';$$

therefore

$$pmA - pnB = pE',$$

and

$$(p + 1)mA - (p + 1)nB = (p + 1)E'.$$

Both which last expressions differ from C by a quantity less than E, the first being less and the second greater than C, and both are of the *form* $mA - nB$, m and n being changed for other numbers.

The common ideas of proportion are grounded entirely upon the false notion that all quantities of the same sort are commensurable. That the supposition is practically correct, if there are any limits to the senses, may be shown, for let any quantity be rejected as imperceptible, then since a quantity can be found as near to B as we please, which is commensurable with A, the difference between B and its approximate commensurable magnitude, may be reduced below the limits of perceptible quantity. Nevertheless, inaccuracy to some extent must infest all *general* conclusions drawn from the supposition that A and B being two magnitudes, whole numbers, m and n, can *always* be found such that $mA - nB = 0$. We have shown that this can be brought as near to the truth as we please, since $mA - nB$ can be made as small as we please. This, however, is not a perfect answer, at least it wants the unanswerable force of all the preceding reasonings in geometry. A definition of proportion should therefore be substituted, which, while it reduces itself, in the case of commensurable quantities to the one already given, is equally applicable to the case of incommensurables. We proceed to examine the definition already given with a view to this object.

Resume the equations

$$mA - nB = 0, \quad \text{or} \quad A = \frac{n}{m} B,$$

$$mP - nQ = 0, \quad \text{or} \quad P = \frac{n}{m} Q.$$

If we take any other expression of the same sort $\frac{n'}{m'}B$ and $\frac{n'}{m'}Q$, it is plain that, according as the arithmetical fraction $\frac{n}{m}$ is greater than, equal to, or less than $\frac{n'}{m'}$, so will $\frac{n}{m}B$ be greater than, equal to, or less than $\frac{n'}{m'}B$, and the same of $\frac{n}{m}Q$ and $\frac{n'}{m'}Q$. Let the symbol

$$\left.\begin{matrix} x \\ z \end{matrix}\right\} >=< \left\{\begin{matrix} y \\ w \end{matrix}\right.$$

be the abbreviation of the following sentence: "when x is greater than y, z is greater than w; when x is equal to y, z is equal to w; when x is less than y, z is less than w." The following conclusions will be evident:

If

$$\left.\begin{matrix} a \\ c \end{matrix}\right\} >=< \left\{\begin{matrix} b \\ d \end{matrix}\right. \quad \text{and} \quad \left.\begin{matrix} a \\ e \end{matrix}\right\} >=< \left\{\begin{matrix} b \\ f, \end{matrix}\right.$$

then

$$\left.\begin{matrix} c \\ e \end{matrix}\right\} >=< \left\{\begin{matrix} d \\ f. \end{matrix}\right. \tag{1}$$

And from the first of these alone it follows that

$$\left.\begin{matrix} ma \\ nc \end{matrix}\right\} >=< \left\{\begin{matrix} mb \\ nd. \end{matrix}\right. \tag{2}$$

We have just noticed the following:

$$\left.\begin{matrix}\dfrac{n}{m}\\[2ex]\dfrac{n}{m}B\end{matrix}\right\} >=< \left\{\begin{matrix}\dfrac{n'}{m'}\\[2ex]\dfrac{n'}{m'}B\end{matrix}\right. \quad\text{and}\quad \left.\begin{matrix}\dfrac{n}{m}\\[2ex]\dfrac{n}{m}Q\end{matrix}\right\} >=< \left\{\begin{matrix}\dfrac{n'}{m'}\\[2ex]\dfrac{n'}{m'}Q.\end{matrix}\right.$$

Therefore (1)

$$\left.\begin{matrix}\dfrac{n}{m}B\\[2ex]\dfrac{n}{m}Q\end{matrix}\right\} >=< \left\{\begin{matrix}\dfrac{n'}{m'}B\\[2ex]\dfrac{n'}{m'}Q\end{matrix}\right. \quad\text{or}\quad \left.\begin{matrix}A\\P\end{matrix}\right\} >=< \left\{\begin{matrix}\dfrac{n'}{m'}B\\[2ex]\dfrac{n'}{m'}Q.\end{matrix}\right.$$

Therefore (2)

$$\left.\begin{matrix}m'A\\m'P\end{matrix}\right\} >=< \left\{\begin{matrix}n'B\\n'Q.\end{matrix}\right.$$

Or, if four magnitudes are proportional, according to the common notion, it follows that the same multiples of the first and third being taken, and also of the second and fourth, the multiple of the first is greater than, equal to, or less than, that of the second, according as that of the third is greater than, equal to, or less than, that of the fourth. This property* necessarily follows from the equations

$$mA - nB = 0,$$
$$mP - nQ = 0,$$

*It would be expressed algebraically by saying that if $mA - nB$ and $mP - nQ$ are nothing for the same values of m and n, they are either both positive or both negative, for every other value of m and n.

but it does not therefore follow that the equations are necessary consequences of the property, since the latter may possibly be true of incommensurable quantities, of which, by definition, the former is not. The existence of this property is Euclid's definition of proportion: he says, let four magnitudes, two and two, of the same kind, *be called proportional*, when, if equi-multiples be taken of the first and third, etc., repeating the property just enunciated. What is lost and gained by adopting Euclid's definition may be very simply stated; the gain is an entire freedom from all the difficulties of incommensurable quantities, and even from the necessity of inquiring into the fact of their existence, and the removal of the inaccuracy attending the supposition that, of two quantities of the same kind, each is a determinate arithmetical fraction of the other; on the other hand, there is no obvious connexion between Euclid's definition and the ordinary and well-established ideas of proportion; the definition itself is made to involve the idea of infinity, since *all possible multiples* of the four quantities enter into it; and lastly, the very existence of the four quantities, called proportional, is matter for subsequent demonstration, since to a beginner it cannot but appear very unlikely that there are any magnitudes which satisfy the definition. The last objection is not very strong, since the learner could read the first proposition of the sixth book immediately after the definition, and would thereby be convinced of the existence of proportionals; the rest may be removed by showing another definition, more in consonance with common ideas, and demonstrating that, if four magnitudes fall under either of these definitions, they fall under the other also. The definition which we propose is as follows: "Four magnitudes, A, B, P, and Q, of which B is of the same kind as A, and Q as P, are said to be proportional, if magnitudes $B+C$ and $Q+R$ can be found *as near as we please* to B and Q, so that A, $B+C$, P and $Q+R$, are proportional according to the common notion, that is, if whole numbers m and n can satisfy the equations

$$mA - n(B+C) = 0,$$
$$mP - n(Q+R) = 0."$$

We have now to show that Euclid's definition follows from the one just given, and also that the last follows from Euclid's, that is, if there are four magnitudes which fall under either definition, they fall under the other also. Let us first suppose that Euclid's definition is true of A, B, P, and Q, so that

$$\left.\begin{matrix} mA \\ mP \end{matrix}\right\} >=< \left\{\begin{matrix} nB \\ nQ. \end{matrix}\right.$$

This being true, it will follow that we can take m and n, so as not only to make $mA-nB$ less than a given magnitude E, which may be as small as we please, but also so that $mP-nQ$ shall at the same time be less than a given magnitude F, however small this last may be. For if not, while m and n are so taken as to make $mA-nB$ less than E (which it has been proved can be done, however small E may be) suppose, if possible, that the same values of m and n will never make

$mP - nQ$ less than some certain quantity F, and let E be the first multiple of F which exceeds Q, and also let E be taken so small that pE shall be less than B, still more then shall $p(mA - nB)$, or $pmA - pnB$ be less than B. But since pF is greater than Q, and $mP - nQ$ is by hypothesis greater than F, still more shall $mpP - npQ$ be greater than Q. We have then, if our last supposition be correct, some value of mp and np, for which

$$mpA - npB \text{ is less than } B,$$

while

$$mpP - npQ \text{ is greater than } Q,$$

or

$$mpA \text{ is less than } (np + 1)B,$$
$$mpP \text{ is greater than } (np + 1)Q,$$

which is contrary to our first hypothesis respecting A, B, P, and Q, that hypothesis being Euclid's definition of proportion, from which if

$$mpA \text{ is less than } (np + 1)B,$$
$$mpP \text{ is } \textit{less} \text{ than } (np + 1)Q.$$

We must therefore conclude that if the four quantities A, B, P, and Q satisfy Euclid's definition of proportion, then m and n may be so taken that $mA - nB$ and $mP - nQ$ shall be as small as we please.

Let

$$mA - nB = E \quad \text{and} \quad E = nC,$$
$$mP - nQ = F \quad \text{and} \quad F = nR.$$

Then

$$mA - n(B + C) = 0,$$
$$mP - n(Q + R) = 0,$$

and since E and F can, by properly assuming m and n, be made as small as we please, much more can the same be done with C and R, consequently we can produce $B+C$ and $Q+R$ as near as we please to B and Q, and proportional to A and P, according to the common arithmetical notion. In the same way it may be proved, that on the same hypothesis $B - C$ and $Q - R$ can be found as near to B and Q as we please, and so that A, $B - C$, P and $Q - R$ are proportional according to the ordinary notion. It only remains to show that if the last-mentioned property be assumed, Euclid's definition of proportion will follow from it. That is, if quantities can be exhibited as near to P and Q as we please, which are proportional to A and B, according to the ordinary notion, it follows that

$$\left.\begin{matrix} mA \\ mP \end{matrix}\right\} >=< \left\{\begin{matrix} nB \\ nQ. \end{matrix}\right.$$

For let $B + C$ and $Q + R$ be two quantities, such that

$$fA - g(B + C) = 0,$$
$$fP - g(Q + R) = 0,$$

in which, by the hypothesis, f and g can be so taken that C and R are as small as we please. We have already shown that in this case (m and n being any numbers whatever) mA is never greater or less than $n(B+C)$, without mP being at the same time the same with regard to $n(Q+R)$. That is, if

$$mA \text{ is greater than } nB+nC,$$

then

$$mP \text{ is greater than } nQ+nR.$$

Take some *given** values for m and n, fulfilling the first condition; then, since C and R may be as small as we please, the same is true of nC and nR; if then

$$mA \text{ is greater than } nB,$$
$$mP \text{ is greater than } nQ.$$

For if not, let $mA = nB + x$, while $mP = nQ - y$, x and y being some definite magnitudes. Then if

$$nB + x > nB + nC,$$
$$nQ - y > nQ + nR,$$

*It is very necessary to recollect that the relations just expressed are true for every value of m and n; and therefore true for any particular case. In this investigation f and g may both be very great in order that C and R may be sufficiently small, and we must suppose them to vary with the values we give to C and R, or rather the limits which we assign to them; but m and n are *given.*

which last equation is evidently impossible; therefore if $mA > nB$, $mP > nQ$. In the same way it may be proved that if $mA < nB$, $mP < nQ$, etc., so that Euclid's definition is shown to be a necessary consequence of the one proposed.

The definition of proportion which we have here given, and the methods by which we have established its identity with the one in use, bear a close analogy to the process used by the ancients, and denominated by the moderns the *method of exhaustions.* We have seen that the common definition of proportion fails in certain cases where the magnitudes are what we have called incommensurable, but at the same time we have shown that though in this case we can never take m and n, so that $mA = nB$, or $mA - nB = 0$, we can nevertheless find m and n, so that mA shall differ from nB by a quantity less than any which we please to assign. We therefore extend the definition of the word proportion, and make it embrace not only those magnitudes which fulfil a given condition, but also others, of which it is impossible that they should fulfil that condition, provided always, that whatever magnitudes we call by the name of proportionals, they must be such as to admit of other magnitudes being taken as near as we please to the first, which are proportional, according to the common arithmetical notion. It is on the same principle that in algebra we admit the existence of such a quantity as $\sqrt{2}$, and use it in the same manner as a definite fraction, although there is no such fraction in reality as, multiplied by itself, will give 2 as the product. But, however small a quantity we may name, we can assign a fraction which, multiplied by itself, shall differ less from 2 than that quantity.

Having established the properties of rectilinear figures, as far as their proportions are concerned, it is necessary to ascertain the properties of curvilinear figures in this respect. And here occurs a difficulty of the same kind as that which met us at the outset, for no rectilinear figure, how small soever its sides may be, or how great soever their number, can be called curvilinear. Nevertheless, it may be shown that in every curve a rectilinear figure may be inscribed, whose area and perimeter shall differ from the area and perimeter of the curve by magnitudes less than any assigned magnitudes. The circle is the only curve whose properties are considered in elementary geometry, and the proposition in question is discussed in all standard treatises on geometry. Indeed, for this or any other curve the proposition is almost self-evident. This being granted, the properties of curvilinear figures are established by help of the following theorem.

If A, B, C, and D are always proportional, and of these, if C and D may be made as near as we please to P and Q, than which they are always both greater or both less, then A, B, P, and Q are proportional.

Let $C = P + P'$, and $D = Q + Q'$, where by hypothesis P' and Q' may be made as small as we please, and A, B, $P + P'$, and $Q + Q'$ are proportionals. If A, B, P, and Q are not proportionals, let P and $Q + R$ be proportional to A and B. Then, since A and B are proportional to $P + P'$

and $Q + Q'$, and also to P and $Q + R$, therefore

$$P + P' : Q + Q' :: P : Q + R$$

in which all the magnitudes are of the same kind. Now let P' and Q' be so taken that Q' is less than R, which may be done, since by hypothesis Q' can be as small as we please. Hence $Q + Q'$ is less than $Q + R$, and therefore $P + P'$ is less than P, which is absurd. In the same way it may be proved that P is not to $Q - R$ in the proportion of A to B, and consequently P is to Q in the proportion of A to B. This theorem, with those which prove that the surfaces, solidities, areas, and lengths, of curve lines and surfaces, may be represented as nearly as we please by the surfaces, etc., of rectilinear figures and solids, form the method of exhaustions.* In this method are the first germs of that theory which, under the name of Fluxions, or the Differential Calculus, contains the principles of all the methods of investigation now employed, whether in pure or mixed mathematics.

---

*For a classical example, see Prop. II. of the twelfth book of Euclid (Simson's edition). Consult also Beman and Smith's *Plane and Solid Geometry* (Ginn & Co., Boston), pp. 144–145, and 190.—*Ed.*

CHAPTER XVII.

APPLICATION OF ALGEBRA TO THE MEASUREMENT OF LINES, ANGLES, PROPORTION OF FIGURES, AND SURFACES.

WE have already defined a measure, and have noticed several instances of magnitudes of one kind being measured by those of another. But the most useful measure, and that with which we are most familiar, is number. We express one line by the number of times which another line is repeated in it, or if the second is not exactly contained in the first, by the greatest number of the second contained in the first, together with the fraction of the second, which will complete the first. Thus, suppose the line A contains B m times, with a remainder which can be formed by dividing B into q parts, and taking p of them. Then B is to A in the proportion of 1 to $m+\frac{p}{q}$, or as q to $mq+p$, and if B be a fixed line, which is used for the comparison of all lines whatsoever, then the line A is $m+\frac{p}{q}$, or $\frac{mq+p}{q}$, if it be understood that for every unit in m, B is to be taken, and also that for $\frac{p}{q}$ the same fraction of B is to be taken that $\frac{p}{q}$ is of unity. In this case B is called the *linear unit.*

But here we suppose that a line B being taken, the ratio of any other line A to B can be expressed by that of the whole numbers $mq+p$ to q, which we have shown in some cases to be impossible. If we take one of these cases, $mA-nB$, though it can never be made equal to nothing, can be made as small as we please, by properly assuming m and n. Let $mA-nB = E$, then $A = \frac{n}{m}B + \frac{E}{m}$, and since $\frac{E}{m}$ can be made as small as we please, A can be represented as nearly as we please by a fraction $\frac{n}{m}$, where B is the linear unit. Hence, in practice an approximation may be found to the value of A, sufficient for any purpose whatever, in the following manner, which will be easily understood by the student who has a tolerable facility in performing the operations of algebra. Let

A contain B, p times with a remainder P,
B contain P, q times with a remainder Q,
P contain Q, r times with a remainder R,

and so on. If the two magnitudes are commensurable, this operation will end by one of the remainders becoming nothing. For, let A and B have a common measure E, then P has the same measure, for P is $A - pB$, of which both A and pB contain E an exact number of times. Again, because B and P contain the common measure E, Q has the same measure, and so on. All the remainders are therefore multiples of E, and if E be the linear unit, are represented by whole numbers. Now, if a whole number be continually diminished by a whole number, it must, if the operation can be continued without end, eventually become nothing. If, therefore, the remainder never disappears, it is a sign that the magnitudes A and B are incommensurable. Nevertheless, approximate whole numbers can be found whose ratio is as near as we please to the ratio of A and B.

From the suppositions above mentioned it appears that

$$
\begin{aligned}
A &= pB + P,^* &(a)\\
B &= qP + Q, &(b)\\
P &= rQ + R, &(c)\\
Q &= sR + S, &(d)\\
R &= tS + T, &(e)\\
&\text{etc., etc.}
\end{aligned}
$$

Substitute in (b) the value of P derived from (a), find Q from the result, and substitute the values of P and Q in (c); find a value of R from the result, and substitute the values of Q and R in (d), and so on, which give the following series of equations:

$$
\begin{aligned}
A &= pB + P,\\
qA &= (pq+1)B - Q,\\
(qr+1)A &= (pqr+p+r)B + R,\\
(qrs+q+s)A &= (pqrs+ps+rs+pq+1)B - S,\\
(qrst+qt+st+qr+1)A &= \\
&(pqrst+pst+rst+pqt+pqr+p+r+t)B + T.
\end{aligned}
$$

*Throughout these investigations the capital letters represent the lines themselves, and not the numbers of units, which represent them, while the small letters are whole numbers, as in the last chapter.

On inspection it will be found that the coefficients of A and B in these equations may be formed by a very simple law. In each a letter is introduced which was not in the preceding one, and every coefficient is formed from the two preceding, by multiplying the one immediately preceding by the new letter, and adding to the product the one which comes before that. Thus the third coefficient of B is $pqr+p+r$; the new letter is r, and the two preceding coefficients are $pq+1$ and p, and $pqr+p+r=(pq+1)r+p$. The remainders enter also with signs alternately positive and negative. Let x, x' and x'' be the nth, $(n+1)^{\text{th}}$, and $(n+2)^{\text{th}}$ numbers of the series p, q, r, etc., and X, X', and X'' the corresponding remainders. Let the corresponding equations be

$$\begin{aligned} a\,A &= b\,B+X, \\ a'\,A &= b'\,B-X', \\ a''A &= b''B+X''. \end{aligned}$$

Here n must be supposed odd, since, were it even, the first equation would be $aA = bB - X$, as will be seen by reference to the equations deduced. Hence, from the law of formation of the coefficients, x'' being the new letter in the last equation,

$$\begin{aligned} a'' &= a'x''+a, \\ b'' &= b'x''+b. \end{aligned}$$

Eliminate x'' from these two, the result of which is $a''b-ab''=ab'-a'b$, the first side of which is the numerator of $\frac{b'}{a'} - \frac{b''}{a''}$, and the second of $\frac{b'}{a'} - \frac{b}{a}$. It appears then that $\frac{b'}{a'}$ is either greater than both $\frac{b}{a}$ and $\frac{b''}{a''}$ or less than both, since $\frac{b'}{a'} - \frac{b''}{a''}$ and $\frac{b'}{a'} - \frac{b}{a}$ will both have the same sign, the numerators being the same and the denominators positive. It may also be proved that $\frac{b''}{a''}$ lies between $\frac{b}{a}$ and $\frac{b'}{a'}$ by means of the following lemma.

The fraction $\frac{m+n}{p+q}$ must lie between $\frac{m}{p}$ and $\frac{n}{q}$; for let $\frac{m}{p}$ be the greater of the two last, or $\frac{m}{p} > \frac{n}{q}$, then $mq > np$, or $\frac{mq}{mp} > \frac{np}{mp}$, or $\frac{q}{p} > \frac{n}{m}$, and $1 + \frac{q}{p} > 1 + \frac{n}{m}$; therefore $\frac{1+\frac{n}{m}}{1+\frac{q}{p}}$ is less than unity, and any fraction multiplied by this is diminished. But

$$\frac{m+n}{p+q} \text{ is } \frac{m}{p} \times \frac{1+\frac{n}{m}}{1+\frac{q}{p}},$$

and is therefore less than $\frac{m}{p}$, the greater of the two. In the same way it may be proved to be greater than $\frac{n}{q}$, the least of the two.

This being premised, since $\frac{b''}{a''} = \frac{b'x''+b}{a'x''+a'}$, it lies between $\frac{b'x''}{a'x''}$ and $\frac{b}{a}$ or between $\frac{b'}{a'}$ and $\frac{b}{a}$.

Call the coefficients of A and B in the series of equations, a_1, a_2, etc., b_1, b_2, etc., and form the series of fractions $\frac{b_1}{a_1}$, $\frac{b_2}{a_2}$, $\frac{b_3}{a_3}$, etc. The two first of these will be $\frac{p}{1}$ and $\frac{pq+1}{q}$, of which the second is the greater, since it is $p+\frac{1}{q}$. Hence by what has been proved $\frac{b_3}{a_3}$ is less than $\frac{b_2}{a_2}$ and greater than $\frac{b_1}{a_1}$; and every fraction is greater or less than the one which comes before it, according as the number of its equation is even or odd. Again, as the numerator of the difference of two successive fractions $\frac{a''}{b''}$ and $\frac{a'}{b'}$, is the same as that of $\frac{a'}{b'}$ and $\frac{a}{b}$, whatever the numerator of the first difference is, the same must be that of the second, third, etc., and of all the rest. But the numerator of the difference of $\frac{p}{1}$ and $\frac{pq+1}{q}$ is 1; therefore either $ab'-a'b$, or $a'b-ab'$, is 1 according as $\frac{b'}{a'}$ or $\frac{b}{a}$ is the greater of the two, that is according as n is odd or even.* Now since the nth and $(n+1)^{\text{th}}$ equations, n being odd, are

$$a\,A = b\,B + X$$

$$\text{and}\quad a'A = b'B - X';$$

*We might say that $ab'-a'b$ is alternately $+1$ and -1; but we wish to avoid the use of the isolated negative sign.

by eliminating A we have

$$(ab' - a'b)B = a'X + aX'$$
$$\text{or} \quad B = a'X + aX'$$

since $ab' - a'b = 1$; and since the remainders decrease and the coefficients increase, $a' > a$ and $X > X'$, whence $2aX < a'X + aX'$, or $2aX' < B$ and $X' < \frac{B}{2a}$; the remainder therefore which comes in the $(n+1)^{\text{th}}$ equation is less than the part of B arising from dividing it into twice as many equal parts as there are units in the nth coefficient of A; and as this number of units may increase to any amount whatever, by carrying the process far enough, $\frac{B}{2a}$ may be made as small as we please, and *à fortiori*, the remainders may be made as small as we please.

The same theorem may be proved in a similar way, if we begin at an even step of the process. Resuming the equations

$$a\ A = b\ B + X,$$
$$a'\ A = b'\ B - X',$$
$$a''A = b''B + X'',$$

we obtain from the second,

$$A = \frac{b'}{a'}B - \frac{X'}{a'};$$

and since $X' < \frac{B}{2a}$, $\frac{X'}{a'} < \frac{B}{2aa'}$, or if B be taken as the linear unit, $\frac{b'}{a'}$ will express the line A with an error less than $\frac{1}{2aa'}$, which last may be made as small as we please by continuing the process.

It is also evident that $\frac{b}{a}$ is too small, while $\frac{b'}{a'}$ is too great; and since X and X' are less than B, $aA < bB+B$, or $\frac{b+1}{a}$ is too great, while $a'A > b'B-B$, or $\frac{b'-1}{a'}$, is too small. Again, $A-\frac{b}{a}B=\frac{X}{A}$ and $\frac{b'}{a'}B-A=\frac{X'}{a'}$. Now $X' < X$ and $a' > a$; whence $\frac{X'}{a'} < \frac{X}{a}$; that is, $\frac{b'}{a'}B$ exceeds A by a less quantity than $\frac{b}{a}B$ falls short of it, so that $\frac{b'}{a'}$ is a nearer representation of A than $\frac{b}{a}$, though on a different side of it.

We have thus shown how to find the representation of a line by means of a linear unit, which is incommensurable with it, to any degree of nearness which we please. This, though little used in practice, is necessary to the theory; and the student will see that the method here followed is nearly the same as that of continued fractions in algebra.*

We now come to the measurement of an angle; and here it must be observed that there are two distinct measures employed, one exclusively in theory, and one in practice.

*See Lagrange's *Elementary Mathematics* (Chicago, 1898), p. 2 et seq.—*Ed.*

The latter is the well-known division of the right angle into 90 equal parts, each of which is one degree; that of the degree into 60 equal parts, each of which is one minute; and of the minute into 60 parts, each of which is one second. On these it is unnecessary to enlarge, as this division is perfectly arbitrary, and no reason can be assigned, as far as theory is concerned, for conceiving the right angle to be so divided. But it is far otherwise with the measure which we come to consider, to which we shall be naturally led by the theorems relating to the circle. Assume any angle, AOB, as the angular unit, and any other angle, AOC (Fig. 11). Let r be the number* of linear units contained in the radius OA, and

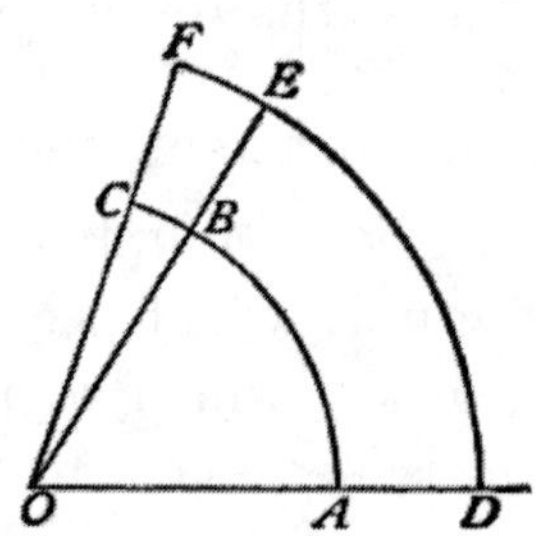

Fig. 11.

t and s the lengths, or number of units contained in the arcs AB and AC. Then since the angles AOB and AOC are proportional to the arcs AB and AC, or to the numbers t and s,

*It must be recollected that the word number means both *whole* and *fractional* number.

we have

$$\text{Angle } AOC \text{ is } \frac{s}{t} \text{ of the angle } AOB;$$

and the angle AOB being the angular unit, the number $\frac{s}{t}$ is that which expresses the angle AOC. This number is the same for the same angle, whatever circle is chosen; in the circle FD the proportion of the arcs DE and DF is the same as that of AB and AC: for since similar arcs of different circles are proportional to their radii,

$$AB : DE :: OA : OD.$$

Also

$$AC : DF :: OA : OD,$$
$$\therefore AB : DE :: AC : DF;$$

therefore the proportion of DF to DE is that of s to t, and $\frac{s}{t}$ is the measure of the angle DOF, DOE being the unit, as before. It only remains to choose the angular unit AOB, and here that angle naturally presents itself, whose arc is equal to the radius in length. This, from what is proved in Geometry, will be the same for all circles, since in two circles, arcs which have the same ratio (in this case that of equality) to their radii, subtend the same angle. Let $t = r$, then $\frac{s}{r}$ is the number corresponding to the angle whose arc is s. This is the number which is always employed in theory as the measure of an angle, and it has the advantage of being independent of all linear units; for suppose s and r to be expressed, for example, in feet, then $12s$ and $12r$ are the numbers of inches in the same lines, and by the common theory of fractions $\frac{s}{r} = \frac{12s}{12r}$. Generally, the alteration of the unit does not affect the number which expresses the *ratio* of two magnitudes. When it is said that the angle $= \frac{\text{arc}}{\text{radius}}$, it is only meant that, *on one particular supposition*, (namely, that the angle 1 is that angle whose arc is equal to the radius,) the number of these *units* in any other angle is found by dividing the number of *linear* units in its arc by the number of *linear* units in the radius. It only remains to give a formula for finding the number of degrees, minutes, and seconds in an angle, whose theoretical measure is given. It is proved in geometry that the ratio of the circumference of a circle to its diameter, or that of half the circumference to its radius, though it cannot be expressed exactly, is between 3.14159265 and 3.14159266. Taking the last of these, which will be more than a sufficient approximation for our purpose, it follows that the radius being r, one-half of the circumference is $r \times 3.14159266$; and one-fourth of the circumference, or the arc of a right angle, is $r \times 1.57079633$. Hence the number of units above described, in a right angle, is $\frac{\text{arc}}{\text{radius}}$, or 1.57079633. And the number of seconds in a right angle is $90 \times 60 \times 60$, or 324000. Hence if ϑ be an angle expressed in units of the first kind, and

A the number of seconds in the same angle, the proportion of A to 324000 will also be that of ϑ to 1.57079633. To understand this, recollect that the proportion of any angle to the right angle is not altered by changing the units in which both are expressed, so that the numbers which express the two for one unit, are proportional to the like numbers for another.

Hence

$$A : 324000 :: \vartheta : 1.57079633;$$

or

$$A = \frac{324000}{1.57079633} \times \vartheta;$$

or

$$A = 206265 \times \vartheta, \text{ very nearly.}$$

Suppose, for example, the number of seconds in the theoretical unit itself is required. Here $\vartheta = 1$ and $A = 206265$; similarly if A be 1, $\vartheta = \frac{1}{206265}$, which is the expression for the angle of one second referred to the other unit. In this way, any angle, whose number of seconds is given, may be expressed in terms of the angle whose arc is equal to the radius, which, for distinction, might be called the *theoretical unit.** This unit is used without exception in analysis; thus,

---

*Also called a *radian.* See Beman and Smith's *Geometry*, p. 192.—*Ed.*

in the formula, for what is called in trigonometry the sine of x, viz.:

$$\sin x = x - \frac{x^3}{2 \cdot 3} + \frac{x^5}{2 \cdot 3 \cdot 4 \cdot 5} - \text{etc.}$$

If x be an angle of one second, it is not 1 which must be substituted for x, but $\frac{1}{206265}$.

The number 3.14159265, etc., is called π, and is the measure, in theoretical units, of two right angles. Also $\frac{\pi}{2}$ is the measure of one right angle; but it must not be confounded, as is frequently done, with 90°. It is true that they stand for the same angle, but on different suppositions with respect to the unit; the unit of the first being very nearly $\frac{206265}{60 \times 60}$ times that of the second.

There are methods of ascertaining the value of one magnitude by means of another, which, though it varies with the first, is not a measure of it, since the increments of the two are not proportional; for example, when, if the first be doubled, the second, though it changes in a definite manner, is not doubled. Such is the connexion between a number and its common logarithm, which latter increases much more slowly than its number; since, while the logarithm changes from 0 to 1, and from 1 to 2, the number changes from 1 to 10, and from 10 to 100, and so on.

Now, of all triangles which have the same angles, the proportions of the sides are the same. If, therefore, any angle CAB be given, and from any points B, B', B'', etc., in one of its sides, and b, b', etc., in the other, perpendiculars be let fall on the remaining side, the triangles BAC, $B'AC'$, bAc, etc., having a right angle in all, and the angle A common, are equiangular; that is, one angle being given, which is not a right angle, the proportions of every right-angled triangle in which that angle occurs are given also; and, *vice versa*, if the proportion, or ratio of any two sides of a right-angled triangle are given, the angles of the triangle are given.

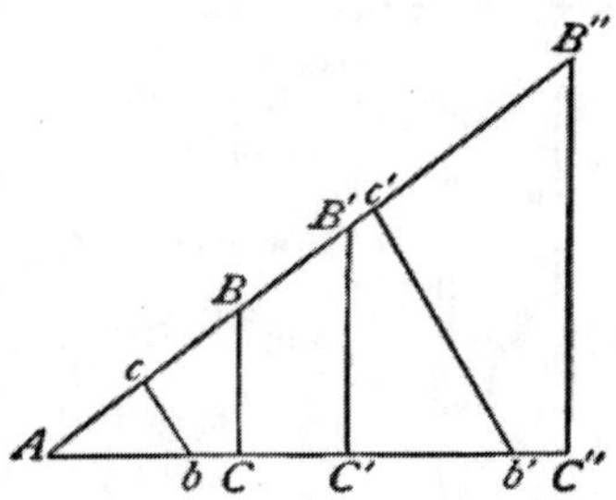

Fig. 12.

To these ratios names are given; and as the ratios themselves are connected with the angles, so that one of either set being given, viz., ratios or angles, all of both are known, their names bear in them the name of the angle to which they are supposed to be referred. Thus, $\frac{BC}{AB}$, or $\frac{\text{side opposite to } A}{\text{hypothenuse}}$, is called the *sine* of A; while $\frac{AC}{AB}$, or $\frac{\text{side opposite to } B}{\text{hypothenuse}}$, or the sine of B, the complement* of A; is called the *cosine* of A. The following table expresses the names which are given to the six ratios, $\frac{BC}{AB}$, $\frac{AC}{AB}$, $\frac{BC}{AC}$, $\frac{AC}{BC}$, $\frac{AB}{AC}$ and $\frac{AB}{BC}$, relatively to both angles, with the abbreviations made use of. The terms *opp.*, *adj.*, and *hyp.*, stand for, *opposite side*, *adjacent side*, and *hypothenuse*, and refer to the angle last mentioned in the table.

THE RATIO	IS THE	BEING	OR	BEING	THESE ARE WRITTEN	
$\frac{BC}{AB}$	sine of A	$\frac{\text{opp.}}{\text{hyp.}}$	cosine of B	$\frac{\text{adj.}}{\text{hyp.}}$	sin A	cos B
$\frac{AC}{AB}$	cosine of A	$\frac{\text{adj.}}{\text{hyp.}}$	sine of B	$\frac{\text{opp.}}{\text{hyp.}}$	cos A	sin B
$\frac{BC}{AC}$	tangent of A	$\frac{\text{opp.}}{\text{adj.}}$	cotangent of B	$\frac{\text{adj.}}{\text{opp.}}$	tan A	cot B
$\frac{AC}{BC}$	cotangent of A	$\frac{\text{adj.}}{\text{opp.}}$	tangent of B	$\frac{\text{opp.}}{\text{adj.}}$	cot A	tan B
$\frac{AB}{AC}$	secant of A	$\frac{\text{hyp.}}{\text{adj.}}$	cosecant of B	$\frac{\text{hyp.}}{\text{opp.}}$	sec A	cosec B
$\frac{AB}{BC}$	cosecant of A	$\frac{\text{hyp.}}{\text{opp.}}$	secant of B	$\frac{\text{hyp.}}{\text{adj.}}$	cosec A	sec B

If all angles be taken, beginning from one minute, and proceeding through $2'$, $3'$, etc., up to $45°$, or $2700'$, and tables be formed by a calculation, the nature of which we cannot explain here, of their sines, cosines, and tangents, or of the logarithms of these, the proportions of every right-angled triangle, one of whose angles is an exact number of minutes, are

*When two angles are together equal to a right angle, each is called the complement of the other. Generally, complement is the name given to one part of a whole relatively to the rest. Thus, 10 being made of 7 and 3, 7 is the *complement* of 3 *to* 10.

registered. We say sines, cosines, and tangents only, because it is evident, from the table above made, that the cosecant, secant, and cotangent of any angle, are the reciprocals of its sine, cosine, and tangent, respectively. Again, the table need only include 45°, instead of the whole right angle, because, the sine of an angle above 45° being the cosine of its complement, which is less than 45°, is already registered. Now, as all rectilinear figures can be divided into triangles, and every triangle is either right-angled, or the sum or difference of two right-angled triangles, a table of this sort is ultimately a register of the proportions of all figures whatsoever. The rules for applying these tables form the subject of trigonometry, which is one of the great branches of the application of algebra to geometry. In a right-angled triangle, whose angles do not contain an exact number of minutes, the proportions may be found from the tables by the method explained in Chapter XI. of this treatise. It must be observed, that the sine, cosine, etc., are not *measures* of their angle; for, though the angle is given when either of them is given, yet, if the angle be increased in any proportion, the sine is not increased in the same proportion. Thus, $\sin 2A$ is not double of $\sin A$.

The measurement of surfaces may be reduced to the measurement of rectangles; since every figure may be divided into triangles, and every triangle is half of a rectangle on the same base and altitude. The superficial unit or quantity of space, in terms of which it is chosen to express all other spaces, is perfectly arbitrary; nevertheless, a common theorem points out the convenience of choosing, as the superficial unit, the square on that line which is chosen as the linear unit. If the sides of a rectangle contain a and b units, the rectangle itself contains ab of the squares described on the unit. This proposition is true, even when a and b are fractional. Let the number of units in the sides be $\frac{m}{n}$ and $\frac{p}{q}$, and take another unit which is $\frac{1}{nq}$ of the first, or is obtained by dividing the first unit into nq parts, and taking one of them. Then, by the proposition just quoted, the square described on the larger unit contains $nq \times nq$ of that described on the smaller. Again, since $\frac{m}{n}$ and $\frac{p}{q}$ are the same fractions as $\frac{mp}{nq}$ and $\frac{np}{nq}$, they are formed by dividing the first unit into nq parts, and taking one of these parts mq and np times; that is, they contain mq and np of the smaller unit; and, therefore, the rectangle contained by them, contains $mq \times np$ of the square described on the smaller unit. But of these there are $nq \times nq$ in the square on the longer unit; and, therefore, $\frac{mq \times np}{nq \times nq}$, or $\frac{mp \times nq}{nq \times nq}$, or $\frac{mp}{nq}$, is the number of the larger squares contained in the rectangle. But $\frac{mp}{nq}$ is the algebraical product of $\frac{m}{n}$ and $\frac{p}{q}$. This proposition is true in the following sense, where the sides of the rectangle are incommensurable with the unit. Whatever the unit may be, we have shown that, for any incommensurable magnitude, we can go on finding b and a, two whole numbers, so that

$\frac{b}{a}$ is too little, and $\frac{b+1}{a}$ too great: until a is as great as we please. Let AB and AC be the sides of a rectangle AK, and let them be incommensurable with the unit M. Let the lines AF and AG, containing $\frac{b}{a}$ and $\frac{b+1}{a}$ units, be respectively less and greater than AC; and let AD and AE, containing $\frac{c}{d}$ and $\frac{c+1}{d}$ units, be respectively less and greater than AB;

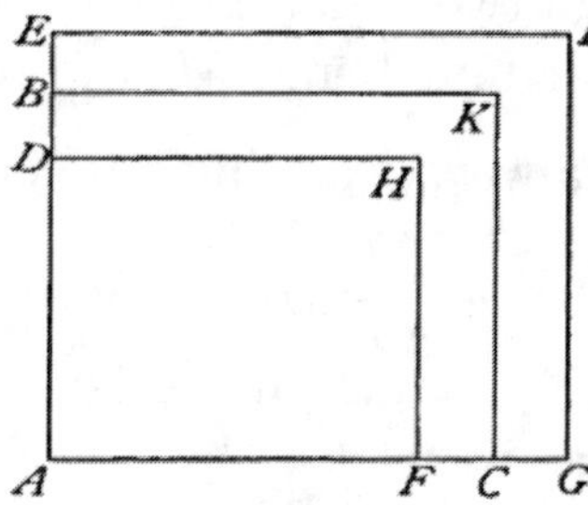

Fig. 13.

and complete the figure. The rectangles AH and AI contain, respectively, $\frac{b}{a} \times \frac{c}{d}$ and $\frac{b+1}{a} \times \frac{c+1}{d}$ square units,* and the first is less than the given rectangle, and the second greater; consequently the given rectangle does not differ from either, so much as they differ from one another. But the difference

* "Square unit" is the abbreviation of "square described on the unit."

of AH and AI is

$$\frac{(b+1)(c+1)}{ad} - \frac{bc}{ad}, \text{ or } \frac{b+c+1}{ad},$$
$$\text{or } \frac{b}{ad} + \frac{c}{ad} + \frac{1}{ad},$$
$$\text{or } \frac{1}{d}\frac{b}{a} + \frac{1}{a}\frac{c}{d} + \frac{1}{ad}.$$

Proceed through two,* four, six, etc., steps of the approximation. The linear unit being M, the results will be such, that $\frac{b}{a}M$ will be always less than AC, but continually approaching to it. Hence $\frac{1}{d}\frac{b}{a}M$ is always less than $\frac{AC}{d}$; and since AC remains the same, and d is a number which may increase as much as we please, by carrying on the approximation, $\frac{AC}{d}$ and *à fortiori* $\frac{1}{d}\frac{b}{a}M$ may be made as small a line as we please; that is, $\frac{1}{d}\frac{b}{a}$ may be made as small as we please, and so may $\frac{1}{a}\frac{c}{d}$ in the same manner. Also $\frac{1}{ad}$ may be made as small as we please; and therefore, also, the sum $\frac{1}{d}\frac{b}{a}+\frac{1}{a}\frac{c}{d}+\frac{1}{ad}$. But this number, when the unit is the square unit, represents the difference of the rectangles AH and AI,

*This is done, because, by proceeding one step at a time, $\frac{b}{a}$ is alternately too little and too great to represent AC; whereas we wish the successive steps to give results always less than AC.

and is greater than the difference of AK and AI; therefore, the approximate fractions which represent AC and AB may be brought so near, that their product shall, as nearly as we please, represent the number of square units in their rectangle.

In precisely the same manner it may be proved, that if the unit of content or solidity be the cube described on the unit of length, the number of cubical units in any rectangular parallelepiped, is the product of the number of linear units in its three sides, whether these numbers be whole or fractional; and in the sense just established, even if they be incommensurable with the unit.

These algebraical relations between the sides and content of a rectangle or parallelepiped were observed by the Greek geometers; but as they had no distinct science of algebra, and a very imperfect system of arithmetic, while, with them, geometry was in an advanced state; instead of applying algebra to geometry, what they knew of the first was by deduction from the last: hence the names which, to this day, are given to aa, aaa, ab, which are called the *square* of a, the *cube* of a, the rectangle of a and b. The student is thus led to imagine that he has proved that square described on the line whose number of units is a, to contain aa square units, because he calls the latter the square of a. He must, however, recollect, that squares in algebra and geometry mean distinct things. It would be much better if he would accustom himself to call aa and aaa the second and third powers of a, by which means the confusion would be avoided. It is, nevertheless, too much to expect that a method of speaking, so commonly received, should ever be changed; all that can be done is, to point out the real connexion of the geometrical and algebraical signification. This, if once thoroughly understood, will prevent any future misconception.

INDEX.